Legislation

Health and safety & environmental

Malcolm Doughton and John Hooper

Australia · Brazil · Japan · Korea · Mexico · Singapore · Spain · United Kingdom · United States

CENGAGE
Learning

Legislation: Health and Safety & Environmental

Malcolm Doughton and John Hooper

Publishing Director: Linden Harris

Commissioning editor: Lucy Mills

Editorial Assistant: Claire Napoli

Project Editor: Alison Cooke

Production Controller: Eyvett Davis

Marketing Executive: Lauren Redwood

Typesetter: S4Carlisle Publishing Services

Cover design: HCT Creative

Text design: Design Deluxe

For product information and technology assistance,
contact **emea.info@cengage.com.**

For permission to use material from this text or product,
and for permission queries,
email **emea.permissions@cengage.com.**

British Library Cataloguing-in-Publication Data
A catalogue record for this book is available from the British Library.

ISBN: 978-1-4080-3988-5

Cengage Learning EMEA
Cheriton House, North Way, Andover, Hampshire, SP10 5BE
United Kingdom

Cengage Learning products are represented in Canada by Nelson Education Ltd.

For your lifelong learning solutions, visit **www.cengage.co.uk**

Purchase your next print book, e-book or e-chapter at **www.cengagebrain.com**

Printed in Malta by Melita Press
1 2 3 4 5 6 7 8 9 10 – 13 12 11

Dedication

This series of study books is dedicated to the memory of Ted Stocks whose original concept, and his publication of the first open learning material specifically for electrical installation courses, forms the basis for these publications. His contribution to training has been an inspiration and formed a solid base for many electricians practising their craft today.

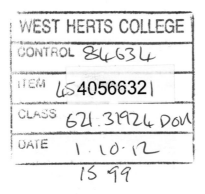

The Electrical Installation Series

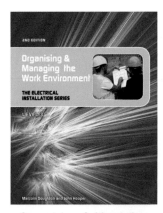

Organising & Managing
the Work Environment

Principles of Design, Installation
and Maintenance

Installing Wiring Systems

Planning and Selection for
Electrical Systems

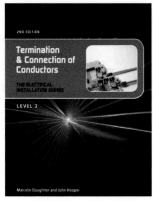

Termination & Connection
of Conductors

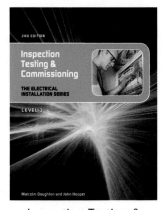

Inspection Testing &
Commissioning

Fault Finding & Diagnosis

Maintaining Electrotechnical
Systems

Contents

Unit two The environment

About the authors

Malcolm Doughton

Malcolm Doughton, I.Eng, MIET, LCG, has experience in all aspects of electrical contracting and has provided training to heavy current electrical engineering to HNC level. He currently provides training on all aspects of electrical installations, inspection, testing, and certification, health and safety, PAT and solar photovoltaic installations. In addition Malcolm provides numerous technical articles and is currently managing director of an electrical consultancy and training company.

John Hooper

John Hooper spent many years teaching a diverse range of electrical and electronic subjects from craft level up to foundation degree level. Subjects taught include: Electrical Technology, Engineering Maths, Instrumentation, P.L.C.s, Digital, Power and Microelectronic Systems. John has also taught various electrical engineering subjects at both Toyota and JCB. Prior to lecturing in further and higher education he had a varied career in both electrical engineering and electrical installations.

Acknowledgements

The authors and publisher would like to thank Chris Cox and Charles Duncan for their considerable contribution in bringing this series of study books to publication. We extend our grateful thanks for their unstinting patience and support throughout this process.

The authors and publisher would also like to thank the following for providing pictures for the book:
Draper Tools Ltd.
JSP Ltd.
Kewtech Corporation Ltd
MK Electric
Rainharvesting Systems Ltd. www.rainharvesting.co.uk
Sealey Power Products Ltd.
Start Traffic Management
Windspire Energy www.WindspireEnergy.com

Study Guide

This studybook has been written and compiled to help you gain the maximum benefit from the material contained in it. You will find prompts for various activities all the way through the studybook. These are designed to help you ensure you have understood the subject and keep you involved with the material.

Where you see "Sid" as you work through the studybook, he is there to help you and the activity 'Sid' is undertaking will indicate what it is you are expected to do next.

 Task

Take a look at your company's safety policy and make a note below of the main points.

1 _____

2 _____

Task A "Task" is an activity that may take you away from the book to do further research either from other material or to complete a practical task. These tasks are included where you are given the opportunity to ask colleagues at work or your tutor at college questions about practical aspects of the subject. There are also tasks where you may be required to use manufacturers' catalogues or access web sites, to look up your answer. These are all important and will help your understanding of the subject.

 Try this

Tick which of the following are required by law to be provided by the employer?

Hand washing facilities
First aid equipment
Kettle

Try this A "Try this" is an opportunity for you to complete an exercise based on what you have just read, or to complete a mathematical problem based on an example given in the text.

Remember

Watch out for new laws regarding Health and Safety and for amendments to the existing laws.

Remember A "Remember" box highlights key information or helpful hints.

Notes

You could find it useful to look in a library for copies of the regulations mentioned in this chapter. Read the appropriate parts and be on the lookout for any amendments or updates to them.

Note "Notes" provide you with useful information and points of reference for further information and material.

RECAP & SELF ASSESSMENT

Circle the correct answers.

1   The Health and Safety at Work etc. Act applies to

a.   employers only

b.   employees only

c.   both employers and employees

d.   safety officers only

2   Under Health and Safety Law which of the following is the responsibility of the employer to provide?

a.   tea making equipment

b.   free laundry service

c.   safety training

d.   weekly safety reports

Recap & Self Assessment At the beginning of all the chapters except the first you will be asked questions to recap what you learned in the previous chapter. At the end of each chapter you will find multichoice questions to test your knowledge of the chapter you have just completed.

The studybook can be divided into Parts, each of which may be suitable as one lesson in the classroom situation. If you are using the studybook for self tuition then try to limit yourself to between 1 hour and 2 hours before you take a break. Try to end each lesson or self study session on a Task, Try this or Self Assessment questions. When you resume your study go over this same piece of work before you start a new topic.

You will find the answers to most of the questions at the back of the book. Answers are not given where they may be found in the preceding text. Before you look at the answers check that you have read and understood the question and written the answer you intended to.

At the back of each unit you will also find a glossary of terms which have been used in the unit.

An "end test" covering all the material in each unit is included so that you can assess your progress.

There may be occasions where topics are repeated in more than one book. This is required by the scheme as each unit must stand alone and can be undertaken in any order. It can be particularly noticeable in health and safety related topics. Where this occurs read the material through and ensure that you know and understand it and attempt any questions contained in the relevant section.

You may need to have available for reference current copies of legislation and guidance material mentioned in this book. Read the appropriate sections of these documents and remember to be on the look-out for any amendments or updates to them.

Your safety is of paramount importance. You are expected to adhere at all times to current regulations, recommendations and guidelines for health and safety.

Unit one

Health and safety

Material contained in this unit covers the knowledge requirement for C&G Unit No. 2351-301 (ELTK 01), and the equivalent EAL unit.

Unit one considers health and safety legislation, the practices and procedures for dealing with health and safety in the work environment and understanding the requirements for identifying and dealing with hazards.

You could find it useful to look in a library or online for copies of the legislation and guidance material mentioned in this unit. Read the appropriate sections and remember to be on the lookout for any amendments or updates to them.

Before you undertake this unit read through the study guide on page vii. If you follow the guide it will enable you to gain the maximum benefit from the material contained in this unit.

1.1 Relevant legislation

At the beginning of all the other chapters in this book you will be asked to complete a revision exercise based on the previous chapter.

Sid at his desk will remind you of this.

For this first chapter exercise make a note of any signs, notices and instructions relating to health, safety, welfare and emergency procedures at your place of work or training centre.

Figure 1.1 *This sign shows the direction to go in an emergency*

_____

_____

_____

LEARNING OBJECTIVES

On completion of this chapter you should be able to:

- Explain employers' and employees' responsibilities under the Health and Safety at Work etc. Act.

- State the procedures that should be followed in reporting health, safety or welfare issues in the workplace.

- State the appropriate responsible persons to whom health, safety and welfare related matters should be reported.

- Recognize other appropriate current legislation and explain employers' and employees' responsibilities.

- State the procedures to be followed in the case of accidents which involve injury.

- State the first aid facilities that must be available in the work area in accordance with health and safety regulations.

Part 1 Health and safety

Note

You could find it useful to look in a library for copies of the legislation and regulations mentioned in this chapter. Read the appropriate parts and be on the lookout for any amendments or updates to them.

Keeping everyone safe at work is the responsibility of both the employer and the employee.

Both employers and employees are required, by law, to observe safe working practices. There are a number of acts of parliament and regulations that govern what employers provide in a workplace and how the employees use this provision.

Figure 1.2 *Cooperating but not working safely!*

Health and Safety at Work etc. Act 1974

This act applies to everyone who is at work.

It sets out what is required of both

- employers and
- employees.

The aim of this act is to improve or maintain the standards of health, safety and welfare of all those at work.

A number of regulations and codes of practice have been introduced under the Health and Safety at Work etc. Act, including:

- Management of Health and Safety at Work Regulations
- The Electricity at Work Regulations
- Manual Handling Operations Regulations
- Control of Substances Hazardous to Health (COSHH) Regulations
- Workplace (Health and Safety and Welfare) Regulations
- Personal Protective Equipment at Work Regulations (PPE)
- Provision and Use of Work Equipment Regulations (PUWER)
- Work at Height Regulations
- Display Screen Equipment at Work Regulations
- Control of Asbestos at Work Regulations.

The Factories Act 1961 can have legal importance in cases that extend back before the Health and Safety at Work etc. Act came into force. This may occur in cases such as industrial noise causing hearing loss or carcinogens which can cause damage to lungs or cancer later in life.

Remember

Watch out for new legislation regarding Health and Safety and for amendments to the existing legislation.

The employer's responsibility

The Management of Health and Safety at Work Regulations requires employers to provide and maintain a working environment for their employees which is, as far as practicable, safe and without risk to health. The 'working environment' applies to all areas to which employees have access. For example, corridors, staircases and fire exits are included, as are gangways and steps.

The 'safe working environment' includes such factors as:

- maintaining a reasonable working temperature and humidity
- adequate ventilation and fume and dust control
- suitable and adequate lighting
- a clean and tidy workplace.

Figure 1.3 *Gangways to be kept clear*

If other employers or contractors are sharing the workplace then the employer is responsible for working with them so that everyone's health and safety is safeguarded.

Other facilities that are required by law include those for washing and sanitation and these must also be suitable for disabled employees.

First aid

The supply of adequate first aid equipment is also the responsibility of the employer. The contents of a first aid box should be based on the employer's assessment of needs required in the company's own situation. All used items should be replaced and those sterile items that have an expiry date must be renewed. Sterile water for eye irrigation may be required and this must be stored in a sealed container and when the seal has been broken the container should not be reused as the water will no longer be sterile.

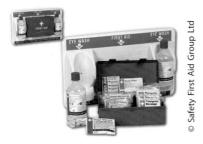

© Safety First Aid Group Ltd

Figure 1.4 *First aid station*

It is important to note where the first aid box is located; this is identified by a white cross on a green background. Also find out who are the appointed first-aiders.

Figure 1.5 *First aid*

Work equipment

Employers are also required to provide and maintain suitable safe tools and equipment for use by their employees. Training in the use of such equipment, where this is necessary, must also be provided. Any information or supervision that may be required is also the employer's responsibility.

Work equipment should be suitable for its intended use and for the conditions in which it is to be used. It should be maintained in a safe condition and inspected periodically to ensure that it remains safe for use. Records should be kept of the inspections carried out on the equipment.

The employer must also ensure that the method of working is safe. Protective equipment, for example machinery guards, safety screens and protective clothing, must be provided where they are required.

The storage, handling and transporting of goods is also the responsibility of the employer. Goods should be stacked on suitable shelves and in a manner that will prevent danger. Some materials, especially chemicals, should be stored in the correct containers and labelled clearly. Heavy items may need to have mechanical handling aids, such as a forklift truck for their safe transportation.

Figure 1.6 *A forklift truck*

Under the Display Screen Equipment at Work Regulations your employer will need to analyze any workstations and assess and reduce risks. Employers have to provide training in the use of a visual display unit (VDU) and workstation so that employees know how best to avoid health problems.

Some common problems, such as aches, pains and tired eyes, can be avoided by having a comfortable workstation and taking frequent short breaks.

Figure 1.7 *The workstation*

Try this

Tick which of the following are required by law to be provided by the employer.

Hand washing facilities
First aid equipment
Kettle
Free laundry
Protective eye wear (where required)
Tea/coffee facilities

Reporting accidents

Employers must have insurance, known as an Employer's Liability Insurance Certificate, in place in case employees get injured or become ill through work. A current insurance certificate must be readily available either as a printed or electronic copy.

Employers must ensure that in the event of an accident, however slight, the details are recorded in an accident book kept at the workplace.

Deaths, major injuries and dangerous occurrences must be reported to the relevant enforcing authority, which may be the local authority Environmental Health Department or the Health and Safety Executive who may wish to investigate further. They may also inspect the accident book at the same time.

Major injuries include those that result in:

- injury from electric shock or burns leading to unconsciousness or requiring resuscitation
- fractured bones
- loss of sight
- hospitalization for more than 24 hours.

Dangerous occurrences that are reportable include:

- the collapse, overturning or failure of machinery such as an excavator or forklift truck

- an electrical short circuit or overload accompanied by fire or explosion which has the potential to cause death or which results in the stoppage of the plant involved for more than 24 hours.

Other work-related diseases, such as asbestosis, which may be caused by working with asbestos dust, must also be reported.

In addition, every accident at work involving an employee due to which the employee is unable to work for three or more consecutive days must be notified to the Environmental Health Department or Health and Safety Executive.

Safety policy

It is a legal requirement for an employer having five or more employees to produce a safety policy and to make it available to all employees. This is usually in the form of a written notice, but it could be stored on a computer; however, it must be available for reference.

The safety policy is produced following a risk assessment which is carried out by the employer.

Figure 1.8 *A safety policy must be available to all employees*

The company's safety policy will include details of any hazards and risks present in the working environment together with the safety procedures which need to be taken in order to protect the health and safety of all persons concerned. This safety policy should be subject to regular review in consultation with the safety representatives.

Task

Take a look at your company's safety policy and make a note below of the main points.

1 _____

2 _____

3 _____

Site visitors

When drawing up their safety policy employers are responsible for ensuring that the safety of visitors to the site is included. Authorized site visitors should be asked to identify themselves and identify, by name, the person they wish to see. They should then sign in the visitor's book and be made aware of the safety procedures they need to follow while they are on site.

The health and safety inspectorate

Employers have many legal duties with regard to health and safety at work, and some of the main responsibilities have already been listed. Breaking health and safety laws can result in prosecution.

The health and safety inspectorate are the Health and Safety Executive (HSE) or the Local Authority Environmental Health Department, who the HSE may appoint to act on their behalf, and they are concerned with the safety and welfare of everybody at work. Whilst they have legal powers to prosecute offenders, it is far more likely that they will issue either an improvement notice or a prohibition notice. Generally it is only where these notices are not acted upon that prosecution will follow.

Improvement notice

An improvement notice can be served on an employer if an HSE inspector is satisfied that a statutory regulation has been contravened.

Once this notice has been served the employer has a specified time (not less than 21 days) in which to take the required action or to appeal to an Industrial Tribunal if they should wish to do so. During the time that an appeal is pending, the improvement notice is suspended and the activity may continue. If the employer fails to comply with an improvement notice within the specified time then prosecution will almost certainly follow.

Prohibition notice

A prohibition notice can have the effect of stopping an activity or practice immediately, without recourse to appeal, if the inspector is satisfied that there is a risk of personal injury. Failure to comply with this notice will almost certainly result in prosecution.

Prosecution carries some significant penalties, for example failure to comply with an improvement or prohibition notice can result in a fine of up to £20 000, or six months' imprisonment.

Higher courts may impose unlimited fines and imprisonment.

Remember

An employer must provide:

- a safe place of work with safe access and exit
- safe equipment and work systems
- a safe working environment including facilities for sanitation, washing and first aid
- safe methods of handling and storing goods
- accident procedure and register
- training and supervision
- a safety policy (for workplaces with five employees or more)
- insurance that covers employees in cases of injury or illness due to work.

 **Try this**

Who does the Health and Safety at Work etc. Act apply to?

_____

What is required of them?

_____

The aim of this Act is to maintain or improve:

_____

The employer must provide and maintain a safe working environment. What areas in your place of work are covered by 'the safe working environment'?

Complete the following list of factors that must be included in 'the safe working environment'.

>   reasonable working temperature
>   reasonable working humidity

>   _____

>   _____

>   _____

>   _____

Employers must also provide and maintain:

>   safe equipment and tools

>   _____

>   _____

>   _____

>   _____

>   _____

On occasions you may be required to wear:

>   eye protection
>   hearing protection
>   head protection

>   _____

>   _____

An employee has an accident and is unable to work for four consecutive days. To whom must the accident be reported and where should it be recorded?

_____

If your company has machinery that fails to comply with legal requirements then the Health and Safety Executive may issue either of the following two notices:

_____

_____

Part 2 Preventing accidents

Employees' responsibilities

An employee can be prosecuted for breaking health and safety legislation.

Employees are required by law to:

- take reasonable care for their own health and safety and not to endanger others
- cooperate with their employer on health and safety procedures
- not interfere with tools, equipment etc. provided for their health, safety and welfare
- correctly use all work items provided in accordance with instructions and training given to them.

Accidents at work

Accidents are **unplanned**.

Accidents do not **'just happen'**.

Accidents are **caused by people**.

An _accident_ is an unexpected or unintentional event which is often harmful.

There are three main causes of accidents:

- people who behave unsafely because of such factors as boredom, horseplay, carelessness or lack of knowledge
- people who have been provided with or have maintained an unsafe environment
- people who have misused or abused a safe environment.

Keeping in mind the causes of accidents we can help prevent accidents by applying the following.

Make the environment as safe as possible: Keep the floors and workbenches clear of rubbish, putting any tools and equipment away when they are not needed. Replace caps on bottles and if any liquids are spilled clean them up immediately.

Figure 1.9 _Replace caps on bottles_

Waste materials should be removed and stored in suitable containers.

Access routes must be kept clear at all times and temporary hazards, such as where a floorboard has been removed, should be adequately guarded.

Do not interfere with machinery: It is extremely dangerous to remove guards. A machine must not be used by someone who has not been suitably trained or has no authority to use it. Long hair should be tied back and loose clothing should not be worn as it may get caught in machinery.

Use the correct tool for the job and make sure it is in a safe condition: Tools which show signs of damage or wear can be potentially dangerous and should not be used until repaired or replaced.

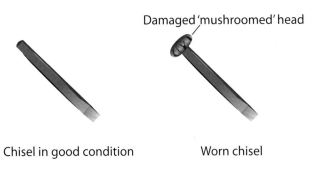

Damaged 'mushroomed' head

Chisel in good condition   Worn chisel

Figure 1.10 *Tool maintenance is important*

Wear protective clothing and other equipment (PPE) if it is considered necessary: For example, eye protection should be worn when grinding metals and hard hats should be worn on all building sites. Protective clothing and equipment should conform, where necessary, to the mandatory standards.

Read the rules, including warning notices and instructional posters, which should be prominently displayed. Other information, instruction and training which may be given is equally important and must be followed.

Figure 1.11 *Wear eye protection*

Safety signs give warning of particular dangers and will show the particular type of protection required for the conditions concerned.

HSE

Figure 1.12 *Wear eye protection*

Employees should work safely and sensibly: Employees should also take reasonable care of their own health and safety and that of others. This means that they will be expected to do their job in a sensible way and they should not run around or fool about.

Figure 1.13 *Work safely and sensibly*

Task

In your place of work find out the following:

1 Who do you have to report damaged tools to?

_____

_____

2 Who issues protective clothing and equipment?

_____

_____

Employees should:

- make sure they are fit for work and not overtired, ill or suffering from the effects of drink or drugs
- always report any health problem they may have to their employer
- look out for hazards in the workplace and report those that they find.

You may be the first to notice a hazard, and if it is ignored it may cause an injury to you or your colleagues. This will include such things as deficiencies in equipment, for example broken ladder rungs, insecure structures and falling objects.

Always report an accident whether it results in injury or not as it may prevent another more serious accident from the same cause.

Any report should be made to an appropriate responsible person. This may be your employer, supervisor or health and safety representative.

Figure 1.14 _Always report potential hazards before they cause injury_

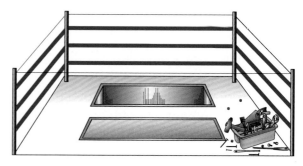

Figure 1.15 _For safety always place a guard around a drop_

Remember

Accidents don't just happen – they are *caused*.

You can help to minimize the number of accidents that occur by:

● learning the safety rules of your workplace and industry and following them

● protecting yourself and the people around you by using the correct equipment, clothing and tools

● keeping a lookout for potential hazards and reporting those you find to the appropriate responsible person

● behaving sensibly and thoughtfully.

Try this: Wordsearch

Find the words from the list below in the wordsearch puzzle:

```
A U T W C G I X A Y T K S I G
E C P W G U A R D S O E F Y J
A N C R V C S W T E L A L H A
A X I I O D K H E U G S O Y T
T H Z J D T Z T R L N P R U E
S I G N S E E O Y O F U U G A
G E G Z L L N C I N J A R W H
X K N V J I I T T N Y J R U E
E Y G F D L A R I I K G W E A
E F S W O L H A A I O G Z K L
M Z X P U H P I X U V N K K T
S C B G G R E N M L V R S J H
I A E O L E G I S L A T I O N
G R F M Z X I N L H A Z A R D
T N N E P S V G Z O X T N L L
```

ACCIDENT	HAZARD	LEGISLATION	REGULATIONS	SIGNS
ACT	HEALTH	POLICY	RULES	TRAINING
GUARDS	INJURY	PROTECTION	SAFE	WELFARE

Task

Find out where the accident book and the first aid box are kept at your place of work.

The accident book is located:

_____

The first aid box is located:

_____

The first aid box contains:

_____

_____

The appointed first-aiders are:

_____

Discuss with your friends and colleagues, at work or at your training centre, any potential hazards which any of you have noticed. Note down any which have not been referred to in this chapter and also what action was taken.

_____

_____

Accident procedure

Following any accident employees need to note the following details in order to fill in the accident record which must be kept in any workplace:

- the date of the accident or dangerous occurrence
- the place where the accident or dangerous occurrence took place
- a brief description of the circumstances
- the name of any injured persons
- the sex of the injured persons
- the age of the injured persons
- their occupations
- the nature of the injuries.

There may be other details required by particular organizations. These may range from witnesses to suggestions for the prevention of further accidents. Major injuries will have to be reported to the appropriate authority.

There are other guides available from the HSE, industry and workplaces that will help in understanding the employees' responsibilities.

Typical site administration

Dabson Electrical Co.

Accident Report Form

This form must be completed in the event of any accident/injury occurring at the above premises or whilst working for the above company.

Date and time of accident:

Date: *Time:*

Place where the accident took place:

Brief description of the circumstances:

Injured person:

Name:

Sex:

Age:

Job title:

Nature of injury sustained:

Your details:

Name:

Job title:

Figure 1.16 *Typical accident report form*

Just as employers are responsible for providing training, information and supervision, so the employees are responsible for carrying out the work in the manner in which they have been trained. Employees are also responsible to the customer in following instructions from their employer, including handing over manufacturers' operating instructions to customers and removing waste materials from the work site.

Remember

These points are important, learn them off by heart.

To prevent an accident:

● Keep the work area clean and tidy – clear up the rubbish!

● Take care near machines – don't interfere with safety guards and be careful not to have long loose hair or clothing.

● Take care of your tools – keep them in good condition.

● Wear protective clothing or equipment when it is necessary.

● Take note of any rules, regulations or safety signs and obey them.

● Take reasonable care for your own health and safety and do not endanger others.

● Look out for hazards and report them – do not assume that this has already been done.

If an accident occurs:

● Report the accident and note down the details required to fill in the accident form.

● Know where the first aid box is located and who is the appointed first-aider.

SELF ASSESSMENT

Circle the correct answers.

1   The Health and Safety at Work etc. Act applies to:

   a.   employers only

   b.   employees only

   c.   both employers and employees

   d.   safety officers only

2   Under Health and Safety Law which of the following is the responsibility of the employer to provide?

   a.   tea making equipment

   b.   free laundry service

   c.   safety training

   d.   weekly safety reports

3   A first aid box is identified by a:

   a.   white cross on a green background

   b.   red cross on a white background

   c.   white cross on a black background

   d.   green cross on a white background

4   A health and safety policy will include details of:

   a.   health and safety advisors' details

   b.   signature of the company secretary

   c.   company aims and objectives

   d.   hazards present in the workplace

5   An employer does not have to provide:

   a.   a guard on an electric saw

   b.   eye protection when chasing brickwork

   c.   hard hats on a construction site

   d.   hand tools for installation work

Electrical safety

RECAP

Before you start work on this chapter, complete the exercise below to ensure that you remember what you learned earlier.

Who does the Health and Safety at Work etc. Act apply to?

_____

How many of the six factors included in the safe working environment can you remember?

_____

_____

_____

_____

_____

_____

If an employee has an accident and is unable to work for four consecutive days where must the accident be recorded and who must the accident be reported to?

_____

LEARNING OBJECTIVES

On completion of this chapter you should be able to:

● State the procedure for producing risk assessments and method statements.

● Describe the procedures for working in accordance with prescribed risk assessments, method statements and safe systems of work.

● Describe methods of circuit isolation.

● Recognize the need to inspect electrical equipment for faults before using it.

Part 1 Electrical accidents

Electrical accidents (or accidents involving the use of electricity) may result in electric shock, fire and burns.

Accidents can occur when equipment or materials are faulty, poorly maintained or misused. Misbehaviour, carelessness, and protective devices and equipment which are either not used or used incorrectly are also contributing factors.

All of these factors apply to electrical accidents.

Figure 1.16

In order to reduce the possibility of accidents when using electricity, strict observance of the applicable legislation, standards and codes of practice is essential.

Legislation intended to prevent electrical accidents

The legislation which we must observe includes:

● The Health and Safety at Work etc. Act
● The Electricity at Work Regulations
● The Electricity Safety, Quality and Continuity Regulations.

BS7671, the Requirements for Electrical Installations, is published by the Institution of Engineering and Technology (IET) and is commonly known as 'the Wiring Regulations'. Whilst these regulations are not statutory they are accepted as standard practice for electrical installation work. They may be referenced in a court of law should a prosecution be brought under the statutory regulations.

The British Standards Institution (BSI) produces other standards and codes of practice. Those which are applicable to electrical installation work should be complied with.

These statutory requirements, standards and codes of practice relate to the manufacture, installation and use of electrical equipment. Staff should be properly qualified, trained and competent before working on electrical equipment.

A competent person is defined by the Construction (Health, Safety & Welfare) Regulations 1996 as:

> Any person who carries out an activity shall possess such **Training, Technical Knowledge or Experience** as may be appropriate, or be supervised by such a person.

Here are some general safety rules to follow. There may be others that are particular to specific organizations or industries and it is important to know which rules apply and ensure they are followed.

General safety when using hand-held portable electrical equipment

Where portable electrical equipment is used in the workplace or on construction sites, accidents can often be prevented by following a few simple rules.

A visual check on cables and plugs, which are particularly liable to damage, can prevent a serious accident.

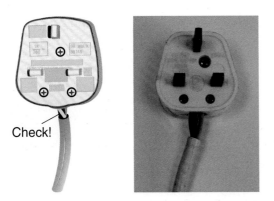

Check!

Figure 1.17 *Visual check of cables and plugs*

Electrical equipment may develop faults which do not affect its operation. The equipment may, however, present a potential hazard. Testing the equipment and implementing repairs helps to ensure accidents are prevented.

Figure 1.18 *Testing equipment before issue*

The exposed metalwork of Class I electrical equipment should be connected to earth by a circuit protective conductor (cpc). However, if the equipment conforms to the appropriate standards it may be classified as 'Class II equipment', often referred to as double insulated. The symbol in Figure 1.19 would be displayed on the case of the equipment and for this Class II equipment no circuit protective conductor is necessary.

© HSE

Figure 1.19 *British Standard symbol for Class II (double insulated) equipment*

Note

The classification of equipment is described in BS 2754:1976 (1999) 'Construction of electrical equipment for protection against electric shock', which is based on IEC Report 536.

Where equipment is used out of doors or in damp or hazardous environments it should have a residual current device (RCD) protecting the circuit with an operating current no greater than 30mA. This means that should a fault develop between line and earth or neutral and earth the circuit would automatically switch off before the fault current could exceed the operating current preventing a dangerous electric shock.

Figure 1.20 *Residual current device (RCD)*

On construction sites, due to the increased shock risk, it is preferable to limit the voltage for handheld portable equipment to 110V ac, known as Reduced Low Voltage. This is supplied through a centre tapped to earth, step-down transformer, as shown in Figure 1.21, which limits the line voltage to 55V above earth potential.

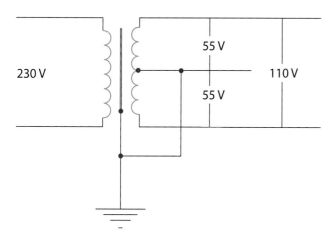

Figure 1.21 *110V centre tapped to earth transformer*

In domestic premises the 13A, BS 1363, plug and socket is used to connect portable equipment to the supply. Whilst this is used in commercial and industrial installations many items of equipment in these locations are connected using a plug and socket conforming to BS EN 60309-2:1992. There are different colours of plug, socket, connector and appliance inlet which are used to identify the voltages they are designed to be used at. So a blue plug can only be connected to a blue socket and so on. So that a plug cannot be connected to the incorrect voltage an interlocking lug is incorporated and is located at different positions for different voltages, as illustrated in Figures 1.23 and 1.24.

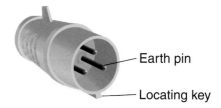

Figure 1.22 *Plug conforming to BS EN 60309*

Figure 1.23 *BS EN 60309 plugs of different voltages*

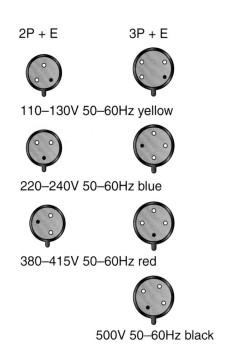

2P + E   3P + E

110–130V 50–60Hz yellow

220–240V 50–60Hz blue

380–415V 50–60Hz red

500V 50–60Hz black

Figure 1.24 *Examples of the position of the earth relative to the interlocking lug for different voltages*

Table 1.1 *Voltage rating and colours of BS EN 60309 plugs and sockets*

Voltages	Colour
50V	white
110–130V	yellow
220–240V	blue
380–415V	red
500V (50–60Hz)	black
>50V (100–300Hz inclusive)	green
>50V (300–500Hz)	green

Try this

Electricity, when used correctly, can be a great help to everybody, but if it is not used safely it can be a killer. Some electrical equipment is made to a standard of insulation so that an earth connection is not required.

If an item of electrical equipment is Class I it must be connected to earth using a _____ _____.

When portable electrical equipment is used on construction sites or in factories _____ voltage is preferred for _____

The letters RCD stand for _____

Explain in your own words what an RCD does.

_____

_____

In industrial locations it is common to use a plug conforming to _____. To ensure that the plug cannot be connected to the incorrect voltage an _____ lug is used which is located at _____ depending upon the voltage.

Part 2 Isolating the supply

In order to work safely on electrical equipment which has previously been energized and put into use, it is essential to ensure the equipment has been isolated before work commences.

Remember

BS7671 defines isolation as:

A function intended to cut off for reasons of safety the supply from all, or a discrete section, of the installation by separating the installation or section from every source of electrical energy.

The requirements of the Electricity at Work Regulations (EAWR) must be followed when isolating the supply.

Before any circuit is to be worked on it should be isolated. Many accidents have occurred where the wrong circuit has been isolated. It is therefore very important that all circuits and isolators are clearly labelled to identify what it is they control. Circuits should be tested to confirm that they are isolated before working on them. It is important that the isolator cuts off the electrical installation, or any part of it, from every source of electrical energy.

The device for cutting off the supply must not only be suitable for operation in both normal and abnormal conditions but must also be placed in an accessible location. Where switches are used as isolators a clear air gap must exist between the contacts so that they cannot accidentally reconnect. They must also be clearly marked so that there is no doubt whether the switch is on or off.

Some typical examples of isolating devices can be seen in Figures 1.25 to 1.29.

Figure 1.25 *Immersion heater (double pole switch)*

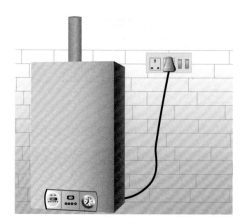

Figure 1.26 *Domestic boiler (plug and socket)*

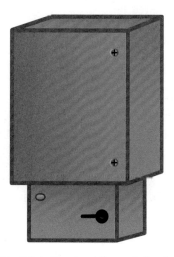

Figure 1.27 *Distribution board (isolator)*

Figure 1.28 *Motor (triple pole isolator)*

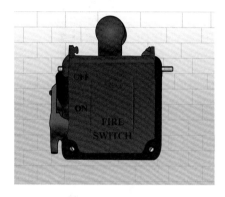

Figure 1.29 *Fireman's switch (isolator)*

Once the equipment has been isolated to work on it there are a number of points that may need consideration including:

- all exposed electrical connections should be tested to see if they are dead
- the test equipment used must be proved to be working correctly before and after use.

It is necessary to post notices at the isolator warning that the equipment is being worked on and the switch or isolator should be **locked** in the **off** position.

If it becomes essential to work on or near live equipment a permit to work may be necessary. This ensures that both the person working on the live equipment and a person in authority know of the risks that are being taken. The important factor is: **do not work on live equipment** without first examining every other possibility. If it is then found necessary to work live, every safety precaution must be considered and taken to prevent danger.

The Electricity at Work Regulations state that work on or near live equipment should be carried out only when:

- it is unreasonable in **all** circumstances for it to be dead
- it is reasonable in **all** circumstances to work on or near the live equipment
- suitable precautions are taken including suitable procedures and protective equipment to prevent injury.

Before work can be started the conductors must be proved 'dead'. Test probes or approved voltage indicators which conform to the Health and Safety Executive Guidance Note GS 38 are recommended for this and they must be checked both before and after testing to ensure that they are working correctly. This is usually carried out with the aid of a proving unit to confirm the approved voltage indicator is functioning before and after testing.

Courtesy of Kewtech

Figure 1.30 *Typical approved voltage indicator and proving unitCourtesy of Kewtech*

Remember

The voltage indicator used to confirm safe isolation must meet the requirements of GS 38.

Where the electrical equipment has its own source of supply, such as batteries, capacitors or generators contained within them these generally cannot be isolated. It is important that in such circumstances all possible precautions are taken to prevent a dangerous situation occurring.

Caution notices should be posted where equipment is isolated warning that work is being carried out on the equipment and that it would be dangerous to reconnect it.

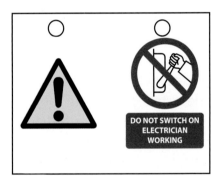

Figure 1.31 *Heavy duty lockout tags*

If there is any live equipment adjacent to isolated equipment then danger notices should be posted to advise these items are live. Once work is completed these notices must be removed.

Permits to work

Permits to work are used for a large number of work activities, many of which have no electrical installation relevance. The purpose of a permit to work scheme is to ensure that activities in areas of risk are controlled and monitored to minimize danger. The risk may be from any source and in particular where the action of others may have a direct influence on the safety of those carrying out the work.

Work carried out in electrical switchrooms, on mechanical plant and processing equipment, aircraft runways, public access areas and secure locations are all likely to be subject to permits to work.

The permit to work system is under the control of a competent person who is charged with the management of all operations and activities requiring such permits. In order to obtain a permit to work, the precise nature of the work involved has to be detailed and a method statement will be required.

The purpose of these details is to ensure that:

● the activity can be assessed
● the implications of the action considered
● an evaluation of the time required can be made.

Other activities being carried out may have implications for the proposed work and ultimately affect both the timing and the extent of the work which can be carried out.

Figure 1.32 *Activities in areas of risk are controlled and monitored to minimize danger*

The permit to work identifies the precise nature of the work to be carried out; any other work which is later found to be necessary should be the subject of a further permit. In certain instances, where work needs to be carried out on, say, a main distribution board, it would involve the authorized person ensuring the appropriate distribution board is isolated and made safe. When this is completed **only** the distribution board which will be isolated and safe to work on is designated on the permit to work.

For work carried out on mains supply networks it is a requirement that, when requesting a permit to work, the network switching details are provided. These are to demonstrate the safe isolation of that part of the network on which work is to take place. Not only must the permit detail the switching arrangements but also the sequence of switching to ensure the minimum loss of supply.

By carefully sequenced switching it is generally possible to isolate a part of the system for maintenance without the loss of supply to the users. Where this is not possible, as in the case of failure or damage, the Electricity Safety, Quality and Continuity Regulations require that the disruption to consumers is kept to a minimum.

The permit to work system is implemented to safeguard everyone involved in the work process and ensures that the statutory requirements are complied with. Ignoring the requirement or exceeding the scope of the permit may endanger yourself or others.

Remember

A permit to work system should be implemented wherever there is a real risk of danger and any actions need to be coordinated and controlled.

Work in such areas must be carried out under the permit and the actions must not extend beyond the scope of the permit issued.

Ignoring the requirements of the scheme or the permit issued can endanger you and others.

Method statement

Having mentioned method statements, it is appropriate to consider their use and content at this time. The purpose of the method statement is to detail precisely what is to be done, how it is to be done, any special requirements or actions and the time anticipated for the work to be carried out. Many companies have a prescribed format for producing a method statement but in general they all contain the above details.

Remember

Isolation is not the operation of a device – it is a complete process which must be undertaken if safe isolation and working are to be achieved.

The basic procedure for isolation requires the following steps to be taken:

- identify the circuit to be isolated
- check the voltage indicating instrument against a known supply or proving unit
- check the circuit is currently live
- isolate the circuit, using the isolation device
- test the circuit to ensure it is dead, using the voltage indicating instrument
- check the voltage indicating instrument against a known supply or proving unit
- secure the isolating device against accidental or unintentional closure.

Providing these steps are followed the circuit can be made safe for work to commence.

Try this: Crossword

Across

3   When terminating cables we must ensure a good one of these (10)

5   One of these boards may contain fuses (12)

7   An isolator must disconnect the installation from all sources of this (6)

10   For a supply on a construction site this should be centre tapped to earth (11)

12   We need one of these for an electric current to flow (7)

13   Damage to 2 down may result in one of these (5)

14   One of these may be required to ensure safe working (6)

Down

1   An industrial one is coloured to indicate the appropriate 4 down (6)

2   This covers conductors to prevent a shock (10)

4   This should be reduced low .... for hand tools on construction sites (7)

6   Before working on a circuit we must carry out safe .... (9)

7   This is used to turn off the light (6)

8   The contact part of a voltage tester (5)

9   This is used to carry current in a circuit (9)

11   A .... current device protects against electric shock (8)

SELF ASSESSMENT

Circle the correct answers.

1   Where equipment is used out of doors in damp environments it should have an RCD protecting the circuit with an operating current no greater than:

   a. 20mA

   b. 25mA

   c. 30mA

   d. 50Ma

2   A BS EN 60309 plug for a voltage of 220–240V is coloured:

   a. blue

   b. red

   c. green

   d. yellow

3   The maximum recommended voltage to earth for hand-held equipment on a construction site is:

   a. 230V

   b. 110V

   c. 55V

   d. 12V

4   When the British Standard symbol, shown in Figure 1.19, is used on equipment this means that it is:

   a. designed for use on 110V

   b. to be earthed at all times

   c. covered by 12 months guarantee

   d. Class II equipment

5   Before any work is carried out on a circuit it is important to confirm that it is:

   a. isolated and locked off

   b. energized and working

   c. connected to the supply

   d. clearly labelled

1.3

Shock, fire and burns

RECAP

Before you start work on this chapter, complete the exercise below to ensure that you remember what you learned earlier.

Portable electrical equipment used on site should be _____ inspected before use.

All Class I electrical equipment should be connected to earth through a
_____.

Class II equipment is often called _____.

The device for cutting off a supply is called an _____.

Test probes should conform to the Health and Safety Executive Guidance Note _____.

A _____ to _____ is a scheme which ensures that activities in areas of risk are controlled and monitored to minimize danger.

LEARNING OBJECTIVES

On completion of this chapter you should be able to:

● State the procedures to be followed in the event of someone receiving an electric shock or burns.

● State procedures to be followed when emergency situations occur in the workplace.

● Identify specific hazards in relation to electric shock, burns, fire and explosions.

● Identify the correct type of fire extinguisher for a particular type of fire.

Part 1 Electric shock

Figure 1.33 *House on fire*

If a person comes into contact with a live conductor whilst also in contact with earth, current will pass through the person's body. This can also happen if a person touches two live conductors at different voltages because the body will complete the circuit between them. In both cases the person concerned will have received an electric shock.

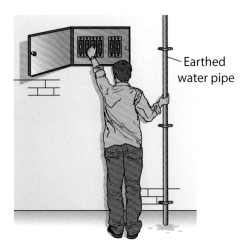

Earthed water pipe

Figure 1.34 *Electric shock between live part and earth*

Electric shock may be caused by the metal casing of electrical equipment becoming 'live'. This may occur if, for example, the circuit protective conductor on an appliance is broken or not connected and the line conductor makes contact with the metal casing, which in turn becomes 'live'. Then anyone touching the casing and other metalwork could become a 'conductor' and suffer an electric shock.

If a voltage, and this can be as low as 50V ac, is applied between two parts of the body it can cause a current to flow that may block the electrical signals between the brain and the muscles. The effect of this could be that the heart stops beating properly and the person is prevented from breathing; it could also cause muscle spasms. These muscle spasms could be strong enough that the person is unable to let go and escape the electric shock. If they are working at height they may fall and if they are working near machinery they may fall into it.

> **Remember**
>
> The voltage is not the only factor. The effects of a current as low as 10–15mA can result in the casualty not being able to let go. Current as low as 30mA can cause breathing difficulties and 50mA can cause the heart to stop beating properly and result in death.

The exact effects will be dependent upon factors such as the voltage, which parts of the body are involved, whether the person is damp, and for how long the current flows.

If you find someone who has received an electric shock your first priority must be to take care not to become a casualty yourself.

For this reason there is a correct procedure that should be followed when dealing with a person who has received an electric shock.

Electric shock procedures

First the connection to the live part must be broken without causing any further injury:

- If possible, isolate, cut off, the electricity supply.
- If this is not possible the casualty will have to be pulled clear **but** only with a dry insulator (dry because water conducts electricity). A dry insulator could be rubber gloves, a dry nylon rope or a wooden pole. Do **not** touch the casualty's bare skin without taking precautions to prevent electric shock.

Once disconnected from the supply:

- If the casualty is not breathing start resuscitation immediately (Figure 1.35) and call for expert help.

Figure 1.35 *Mouth to mouth resuscitation*

- If the casualty is unconscious, but breathing, place them in the recovery position (Figure 1.36) and get help. The injured person may also be burned or bleeding as a result of the shock or fall and may require further medical aid.

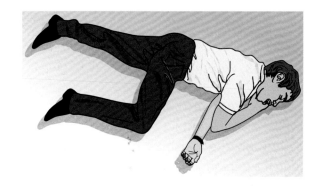

Figure 1.36 *The recovery position*

Remember

Your first priority when someone has received an electric shock is to take care not to become a casualty yourself. Learn and follow the correct procedure!

Electrical burns

Higher currents can create arcs that may cause serious burns. The electrical current passes through the body and heats the tissue along the length of the current flow and this can result in deep burns. With high current the voltage may be very low, so the electric shock potential is slight.

A good example of how a high-current, low-voltage electrical burn can be caused is the misuse of a lead acid battery as used in a car. The voltage is only 12V dc, but if the battery is shorted out the current can be over 100A. A spanner carelessly laid down or dropped across the terminals of such a battery will cause an arc. This produces considerable heat and may cause very serious burns.

Faulty, overloaded, or incorrectly maintained electrical equipment can get hot in normal operation. A faulty lead acid car battery could explode if it is shorted out.

Thermal burns can be the result of a person getting too near hot surfaces or being near to an electrical explosion.

If you are working in an explosive atmosphere, such as in a paint spray booth or near fuel tanks, a single extra low voltage torch battery can generate a spark powerful enough to cause a fire or explosion.

If you or a colleague receives an electrical burn get expert medical help immediately.

Fire hazards

In order for a fire to start there must be fuel (a flammable substance), air (oxygen) and a source of ignition (heat). If all three are not available then a fire will not start. If a fire has started, then by removing any one of the three the fire will be extinguished.

Enough heat can be produced during initial combustion to maintain or accelerate the fire, resulting in the fire rapidly spreading.

Fires can be started in numerous ways such as people smoking carelessly, friction heat, sparks, naked flames, fuel leaks and faults or failures in equipment resulting in overheating and chemical reactions.

If you eliminate any one of these elements then the fire will go out

Figure 1.37 *The fire requirements*

For example the use of a foam fire extinguisher cuts off the supply of oxygen. Alternatively, if there is no more fuel or combustible material, then a fire will go out.

Figure 1.38 *Enforce the no smoking rule in areas where it could start a fire*

Preventing fires

There are precautions that we can take to prevent fires from starting, many of which are common sense. We should take care to store combustible materials away from heat sources, and all areas should be kept free from combustible rubbish and dust.

Equipment which could cause a fire should be regularly maintained and serviced, including electrical equipment and associated cables, plugs and flexes. Fuel pipes should be checked for leaks and flammable materials should be suitably stored at controlled temperatures.

The Fire Precautions (Workplace) Regulations 1997 cover places of work where one or more persons are employed.

The employer in a workplace with more than five workers must produce a written fire risk assessment in order to safeguard the safety of employees in the case of a fire.

Where appropriate the following must be provided:

- emergency exit routes
- final emergency exit doors opening outwards and not sliding or revolving doors
- emergency lighting where necessary covering the exit routes
- fire alarms, fire detectors and fire-fighting equipment where necessary
- fire exit signs (pictograph type).

Employees must be trained in fire safety and follow the written risk assessment.

If a fire occurs ...

Make sure that you understand the hazards involved and know what to do in the event of a fire.

Places of work are required to have a fire procedure and if a fire occurs this should be followed immediately.

This will generally involve:

- raising the alarm
- calling the fire service, giving them your name, the address of the premises you are calling from and any injuries already caused that you know of
- using the correct fire fighting equipment for the particular type of fire where appropriate
- shutting down equipment if possible
- clearing personnel from the area to the fire assembly point
- shutting all doors when evacuating the premises to prevent the spread of fire and smoke and to limit the extent of damage to property
- reporting to the supervisor at the prearranged assembly point.

The fire drill procedure that is displayed in your place of work must always be followed. Make sure you are aware of the emergency procedures which cover evacuation in the event of other emergencies, for example in the case of an explosion, contaminated fumes or terrorist activity.

Fire safety regulations

Regulations covering fire safety include:

Health and Safety at Work etc. Act 1974

Fire Precautions (Workplace) Regulations 1997

Building Regulations 2010

Fire and Rescue Services Act 2004.

Task

Find out the fire drill procedure either at your place of work or your college/training centre. Make a note of the procedure below.

_____

_____

_____

_____

_____

_____

Part 2 Fire safety equipment

It is important to find out the location of all available fire fighting equipment and the types of fire for which it can be used.

Fire extinguishers

Before using a fire extinguisher ensure that you are between the fire and the exit route. Check the label on the fire extinguisher and ensure it is appropriate for the type of fire and safe to use. Do not use water or foam based fire extinguishers on an electrical fire as you may get a fatal electrical shock. Most fire extinguishers require you to remove the safety catch or pin before use.

All British-made fire extinguishers are colour-coded to indicate their particular purposes; Figure 1.39 shows the different types and main uses. In the UK colour identification panels are placed on or above the operating instructions. Fire extinguishers with full body colour coding may still be found and can continue to be used until they need replacing. Halon extinguishers, colour code green, have been withdrawn from service due to the adverse effects they had on the environment.

Type of fire	Water (red panel)	Foam (beige panel)	Dry powder (blue panel)	CO$_2$ gas (black)
Class A Paper, wood, textiles	✓	✓	✓	
Class B Flammable liquids such as oil, paint, petrol, paraffin, grease		✓	✓	✓
Class C Flammable gases such as LPG, butane, propane, methane			✓	✓
Electrical hazards				✓

Pictures supplied by Draper Tools Limited and Sealey Power Products Ltd

Figure 1.39 *Fire extinguisher colour codes and their use.*

Fire blankets

Fire-resistant blankets are used to cut off the air supply to extinguish the fire. These are generally used on fat or oil fires, such as when a deep fat fryer catches fire. They can also be used to smother clothing fires.

Figure 1.40 *Fire blanket safety sign*

Burn injuries

If you are to deal with minor burns you should first wash your own hands. Do not remove any clothing that may be sticking to the burn. Initially cool the burn under running water for 10 minutes, or until the burning sensation has stopped, and then apply a non-adhesive sterile dressing. If the burn is serious, get expert medical help immediately.

Emergencies

An emergency is any event which requires immediate action and in the event of an emergency your priority is to limit injury to persons before limiting damage to property.

You should:

- raise the alarm
- notify the professional emergency services
- suspend work immediately
- isolate equipment from its power source if it is safe to do so
- then proceed in accordance with safety procedures to a recognized assembly point.

Remember

It is important to know how to treat a person who has received an electric shock, without causing further accidents and injury.

It is better to prevent a fire occurring if possible, but should fire break out then the correct extinguisher must be used.

Keeping safe means that you must:

- stay alert
- be aware of how to protect yourself and others
- know what to do in an emergency
- report all hazards you cannot cope with yourself.

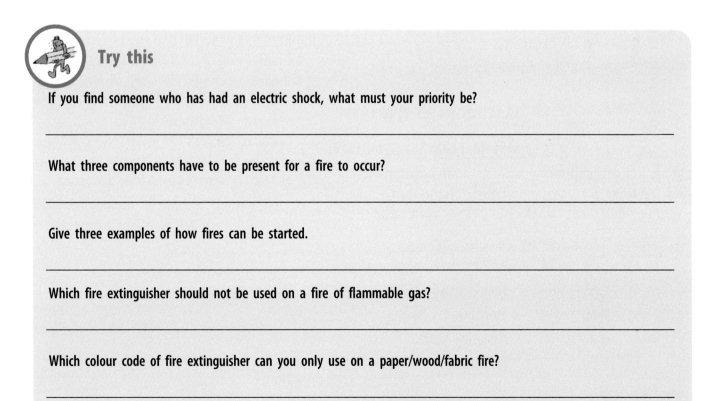

Try this

If you find someone who has had an electric shock, what must your priority be?

_____

What three components have to be present for a fire to occur?

_____

Give three examples of how fires can be started.

_____

Which fire extinguisher should not be used on a fire of flammable gas?

_____

Which colour code of fire extinguisher can you only use on a paper/wood/fabric fire?

_____

Task

See how many examples of fire-fighting equipment you can find.
Note the type of equipment, its location and the use to which it can be applied.

e.g	extinguisher with red label (water)	Site hut	wood/fabric
—	_____	_____	_____
—	_____	_____	_____
—	_____	_____	_____
—	_____	_____	_____
—	_____	_____	_____

SELF ASSESSMENT

1 The first action to be taken when finding someone suffering from an electric shock is:

   a. phone for an ambulance
   b. disconnect the supply
   c. turn them over on their back
   d. run for help

2 The three elements that all fires require are:

   a. fuel, air and a source of ignition
   b. fuel, air and oxygen
   c. fuel, a source of ignition and paper
   d. heat, air and oxygen

3 Which of the following fire extinguishers is correct for the type of fire identified?

	Extinguisher	Fires involving
a.	foam	electricity
b.	water	flammable liquids
c.	foam	propane
d.	dry powder	LPG

4 If a person touches a live conductor and earth the likely outcome is that they will:

   a. suffer no ill effects
   b. receive a shock
   c. not conduct current
   d. become an insulator

5 A colleague receives a minor burn to his hand, your first action for the casualty is to:

   a. send your colleague home
   b. flush the burn with cold water
   c. cover the burn with a dry dressing
   d. remove any burnt material from the wound

The working environment

1.4

RECAP

Before you start work on this chapter, complete the exercise below to ensure that you remember what you learned earlier.

When someone receives an electric shock the first thing to do is break the _____ with the supply without causing any further injury.

If casualties have to be pulled clear then only a _____ insulator should be used.

In order for a fire to start there must be three elements which are:

_____

_____

_____

Do not use a _____ or _____ fire extinguisher on an electrical fire.

Which fire fighting equipment should you use on a chip pan fire? _____

LEARNING OBJECTIVES

On completion of this chapter you should be able to:

- Recognize the need to take particular environmental conditions into consideration before starting work.

- State particular health and safety risks and requirements of current health and safety legislation which may be present when preparing and planning electrotechnical work operations.

- Recognize different types and conditions of worksite.

- Recognize the need for the correct storage of components and equipment.

- Carry out basic assessments of possible risks on site.

- Recognize the requirements for restoring the worksite.

- Recognize the need to take additional precautions in situations where chemicals are present.

- Define the term 'hazard'.

- Identify possible hazards and explain the practices and procedures for dealing with them.

- Identify hazardous substance warning signs.

- Identify situations where asbestos may be encountered and the procedures required for dealing with the suspected presence of asbestos in the workplace.

The electrical industry requires work to be undertaken in a wide range of work locations, the different environments encountered on these various worksites may introduce additional hazards.

Figure 1.41 *Working on site*

Figure 1.42 *Working on industrial sites can involve varying work conditions*

Part 1 Preparing for work on site

Relevant legislation

In this chapter we are going to consider the requirements of the following regulations in particular:

- The Health and Safety at Work Regulations 1992
- Workplace (Health, Safety and Welfare) Regulations 1992
- Provision and Use of Work Equipment Regulations 1988 (PUWER)
- Control of Substances Hazardous to Health (COSHH) Regulations
- Control of Asbestos at Work Regulations.

Before any work can be started your employer should make a risk assessment. Where tasks are routinely carried out your company should have a risk assessment for each task which will be applicable to most locations. However, there is still a need to carry out a risk assessment at each worksite to confirm the arrangements are satisfactory.

Consideration should be given to:

- the work environment
- tools and materials that may be required
- personal protective equipment
- substances used, found or created in the work environment.

The work environment is the subject of the Workplace (Health, Safety and Welfare) Regulations 1992.

As we established in Chapter 1.1, these regulations place a duty on employers to ensure that, as far as is reasonably practical, the workplace meets the health, safety and welfare needs of their employees.

In general employers are required under '**health**' in the work environment to consider the following:

Ventilation: Workplaces should be adequately ventilated and dust, fumes or vapours need to be controlled.

Temperature: Some indoor workplaces could cause 'heat stress', such as in a foundry or laundry. The effects of heat could be controlled by increasing the ventilation or wearing appropriate clothing.

Outdoor workplaces in the winter, or areas such as cold stores, may give rise to 'cold stress'. Again wearing appropriate clothing can reduce the risk to health.

In both the above situations, consideration should also be given to the period for which employees will be exposed to the environment.

Lighting: Lighting should be provided so that employees can work and move about safely. Where a sudden loss of light would create a risk then emergency lighting which operates automatically should be provided.

Under '**safety**' the following are important:

Maintenance: Maintenance is necessary to keep the equipment, devices and systems in safe and efficient working order.

Floors and access areas: All areas that are pedestrian or vehicle routes should be maintained to enable people and vehicles to circulate safely. They should be kept clear of obstructions and should not have holes in them or be uneven or slippery, which could cause a person to trip or slip.

Under '**welfare**':

Washing facilities, sanitary conveniences and a supply of drinking water should be provided.

Employers are also required to provide adequate health and safety training and comprehensive and relevant health and safety information.

General wear for work

In all work situations and locations suitable clothing should be worn. This clothing should not introduce additional hazards – for example torn or loose clothing could get caught in equipment, flimsy footwear will not protect against hard and sharp objects and it could be advantageous on some jobs to wear overalls.

Equipment and materials

Generally, any equipment which is provided for use by employees in the course of their work is controlled under the Provision and Use of Work Equipment Regulations 1988. This covers all work equipment including items such as photocopiers, ladders, electric drills and circular saws. These items must be suitable for use and maintained in a safe condition.

Preparing the worksite

Electrical installation work is carried out in many different locations: domestic, agricultural, commercial and industrial. These premises may be in the course of construction, being extended or modified and could be vacant or occupied.

The electrician's working environment will also vary greatly including:

● a comfortable home
● damp and noxious atmospheres
● heat, such as in a boiler room
● cold, such as in a cold store, or working outside in wintry conditions.

Electrical installation work involves working both inside and outside of buildings. In some cases the wiring will be run on the surface and in others the cables may be concealed under floors, in roofs and in building voids. Some wiring

Figure 1.43   *All the comforts of the home?*

systems must be installed at height, whilst others are buried in trenches.

The type of installation will depend on the construction of the building, its purpose and whether it is a permanent or temporary structure.

The construction of buildings may involve the use of wood, brick, steel, concrete and plasterboard. Each of these materials has different properties which have to be recognized so that fixings and openings can be made effectively with minimum damage.

Risks

Before any work is undertaken the site must be inspected for any potential dangers.

These could include:

- insecure structures
- inadequate lighting
- the risk of falling or being hit by falling or moving objects
- risk of drowning
- dangerous or unhealthy atmospheres
- steam, smoke or vapours.

Wherever possible any risks should be removed, or access to the danger area prevented by barriers and warning notices. Consideration must also be given to the provision, wearing and use of the appropriate personal protective equipment (PPE) which includes protective clothing and equipment.

> **Note**
>
> Guidance on the requirements for PPE can be found in the HSE Guidance INDG174(rev1), a short guide to the Personal Protective Equipment at Work Regulations 1992.

Supply services

Services such as electricity, gas, water, steam or compressed air are often present on site and the location of these must be established. The

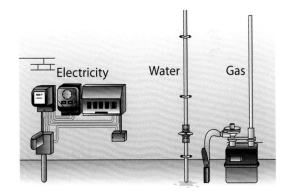

Figure 1.45 *Identify supply services*

Figure 1.44 *An insecure structure?*

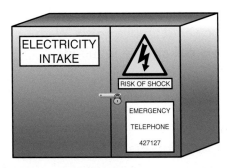

Figure 1.46 *The electricity intake for the site has been labelled and sealed*

metallic gas and water installation pipes will need to be bonded and any deficiencies should be recorded and appropriate action taken. It must also be determined whether the electrical supply service is in a safe condition and adequate for any future additional load likely to be connected.

Task

List two different worksites where each of the following would be a hazard:

Dampness

_____   _____

Corrosive vapours

_____   _____

Drowning

_____   _____

Extreme heat

_____   _____

Extreme cold

_____   _____

Working at heights

Where work at height is necessary suitable access equipment must be provided in the form of steps, trestles, ladders, scaffolding, platforms or powered mobile hoists. Warning notices and barriers will also be required.

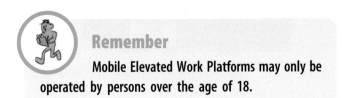

Remember

Mobile Elevated Work Platforms may only be operated by persons over the age of 18.

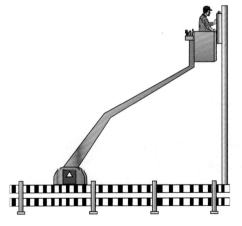

Figure 1.47 *Where risks are involved warning notices and barriers should be erected*

Customer disruption

Where work is to be undertaken in occupied premises it can be the cause of considerable disruption to the occupiers and their visitors or clients.

Figure 1.48 *Minimizing the disruption?*

For example, it may be necessary to:

* disconnect power supplies to businesses or to essential facilities such as in hospitals. In such cases arrangements would have to be made for work to be carried out when least disruption would be caused.

* erect access platforms in a factory to install lighting equipment

* run cables out on a busy shop floor prior to pulling them into position.

In cases like these, arrangements would have to be made for the work to be done either outside of normal hours or when least disruption would be caused.

Remember

An employer must:

* **assess the risk before a job is started**
* **provide the right equipment for the job**
* **give you training in working at heights.**

Task

What preparations would you need to make if your company has been asked to provide extra power points to supply additional milking stalls on a local farm. The cattle will continue to be milked in existing stalls in the same area.

List potential dangers you may encounter on the following sites:

Nursery school   Swimming pool

Try this

Give two possible considerations to be made before starting work in the following environments:

domestic

_____

_____

agricultural

_____

_____

office

_____

_____

shop

_____

_____

construction site

_____

_____

hospital

_____

_____

Dangerous risks must either be removed or access to the danger area prevented by _____ and _____.

Supply services such as _____,_____, water, steam or _____ are often present on site and their position established. The location of the gas and water installation pipes are necessary as they may need to be _____.

Tools

Tools for fixings

When working on site different materials will require different tools for fixings.

Table 1.2 shows some of the more common materials together with appropriate fixings and suitable tools for each type of material.

© Pictures supplied by Draper Tools Ltd

Figure 1.49 *Rotary hammer*

Table 1.2

Material types and properties	Fixings	Suitable tools
Brickwork Common or facing bricks are made from clay. They are fairly easy to drill and have good fixing properties. Engineering bricks are harder and fairly difficult to drill.	Screw fixings Fibre or plastic plugs Masonry nails	Hand drill Electric drill Rotary hammer drill Hammer
Concrete Made from cement and aggregate in dense and lightweight forms as for concrete blocks. **Concrete blocks** Lightweight (clinker) concrete blocks are easy to drill and have fair fixing properties. Dense (limestone) concrete blocks are fairly easy to drill and have good fixing properties.	Screw fixings Fibre or plastic plugs Bolts – several different methods Resin anchors Stud anchors	Hand Drill Electric drill Rotary hammer drill
Steel Used in the construction of many commercial and industrial buildings and has high mechanical strength, although corrosion must be considered. Fixings are generally good and easily made, where threading of steelwork is involved these are time consuming.	Snap on fixings, girder clips cartridge shot fired fixings Drilled and tapped threaded holes with bolts or screws.	Hammer, screwdriver and spanner Cartridge powered fixing tool
Plasterboard Lath and plaster Hardboard, plywood, fibre board and chipboard Difficult to get good fixings	Spring toggle fasteners Gravity toggle Cavity sleeve Cavity wall anchor Dry lining boxes	Drill Screwdriver
Wood Easy to drill Good fixing properties	Wood screws Nails	Drill Screwdriver Hammer

Power tools, such as drills, both mains and battery operated, are commonly used to make fixings into building materials. Making fixings into steel, masonry, concrete and brickwork can be achieved much quicker using a cartridge tool. These tools do have inherent dangers for the installer and others and so they must **only be used by trained operators**.

Materials

When equipment and materials such as switchgear, conduit, trunking, cable and wiring accessories are brought onto the site provision must be made for their safe and secure storage. Where larger quantities of materials are needed for the job it is important to have a system in place to ensure the right quantity of each material is available at each stage of the work or the job could become idle.

Once the materials have been delivered, checked and any omissions or deficiencies recorded, it is important to protect them from damage or theft. A site hut, or a room on the site which can be locked, may be required as a store for this purpose. This store must be laid out so that the materials and equipment are not damaged.

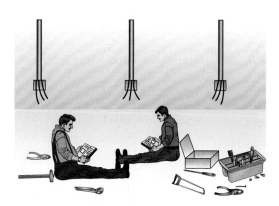

Figure 1.50 *Lack of material and the job becomes idle!*

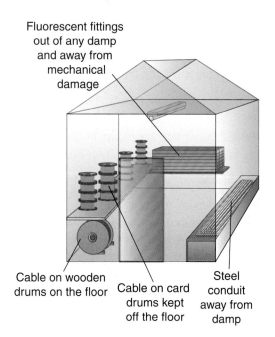

Fluorescent fittings out of any damp and away from mechanical damage

Cable on wooden drums on the floor

Cable on card drums kept off the floor

Steel conduit away from damp

Figure 1.52 *Typical storage of electrical equipment in a site hut*

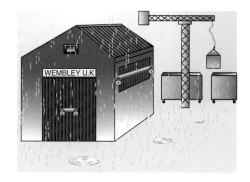

Figure 1.51 *Materials need protection from theft and adverse environmental conditions*

The atmosphere in the store should not be allowed to become damp to ensure the equipment remains in good condition prior to installation on site. Steel conduit could rust and cable drums disintegrate if they are kept in damp conditions. Fluorescent fittings need to be stored to prevent any possible mechanical damage.

The protection of equipment does not end with just a recognized store as once it has been installed the equipment can still become damaged. It is sometimes wise to remove the inside assemblies of equipment and only fit the case while other trades are working in the area. This of course means that the inside assemblies then require even better storage facilities as they no longer have their cases for protection.

Within the storage area there may also be space for the tools and plant that is being used on the site.

Remember

Product quality is preserved and protected by using the correct handling techniques. It is important to ensure that:

● stock is unpacked using the correct techniques and equipment
● all packing is removed and disposed of promptly and in the correct manner
● discrepancies and/or damaged stock are set aside to be dealt with correctly
● in adverse conditions precautions are taken to prevent damage.

Instructions

The instructions regarding how the work is to be carried out will depend on the size and complexity of the contract. These may vary between simple verbal instructions and complicated written specifications, site plans and programmes involving the cooperation of customers, customers' agents and other site trades. For complex contracts a programme will be drawn up to show which trade is working on the site when, so that different trades do not get in each other's way.

Any change in the work programme, for example a delay to one of the trades, may result in a knock-on effect to all the other work. If problems are experienced in keeping to the programme, perhaps due to sickness or bad weather, the

Figure 1.53 *Do not get in each other's way!*

site agent should be notified promptly so that adjustments can be made and any adverse effects minimized.

Hazards in the work environment

A hazard is something that could potentially be harmful to a person's life, health, property, or the environment.

In cold or wet situations care must be taken as surfaces may become slippery, particularly when carrying heavy or bulky objects which may create additional hazards if dropped. When working in damp situations you are likely to feel the cold more and if you have cold or wet hands you may not be able to grip safely when holding tools or climbing ladders.

Wet working, for example when cleaning equipment or using wet cement, can cause dermatitis and this may not become apparent for many years.

So if control measures are taken, and this includes using the right equipment in the correct manner and wearing PPE when the occasion demands, then your health need not be harmed.

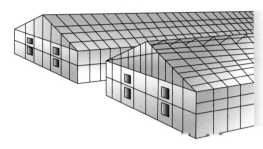

Figure 1.54 *Working in cold or wet situations can cause problems*

Task

Your employer should have taken into account any hazards introduced by adverse working environments. Check any safety notices and policies and make a note of the hazards that have been identified together with the precautions required to overcome or reduce the hazard.

Part 2 Chemical hazards

Employers are required, by law, to take measures to control exposure to hazardous substances and protect their employees' health.

In the case of a chemical, covered by the **Control of Substances Hazardous to Health Regulations (COSHH)**, your employer must consider whether the process will produce gas, fumes, vapour or dust and if the substance is harmful to inhale or will harm your skin. Every year thousands of workers contract lung diseases, such as asthma or cancer, or their skin is affected by dermatitis. Other substances such as cutting compounds, adhesives, lubricants and paint can be harmful. Solvent-based products can give off flammable vapours and dust from wood or flour can explode if ignited. Employers must evaluate all the risks to health and implement action to remove or reduce the risks.

Figure 1.55 *Chemicals are used in electro-plating baths*

Figure 1.56 *Flour stores are dusty environments*

Figure 1.57 *Over time concrete dust may cause lung damage*

Suppliers of hazardous chemicals have to comply with the **Chemicals (hazard information and packaging for supply) 2009 regulations (**CHIP**)** under which they have to provide safety data sheets and ensure that their products are labelled and packaged correctly.

International symbols have replaced the earlier European ones. It is essential that the hazard statement on the packaging and the safety data sheet from the supplier are read.

Figure 1.58 *New international symbols (HSE COSHH)*

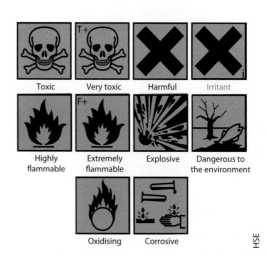

Figure 1.59 *European symbols that you may still see (HSE COSHH)*

Asbestos

The Control of Asbestos at Work Regulations 2006 applies to all work with asbestos. In particular it applies to any work which disturbs material containing asbestos. There are three main types of asbestos – blue, brown and white – but the term 'asbestos' also covers materials which are mixtures containing any of these fibrous silicates. Any employee whose work is likely to expose them to asbestos should be given training in asbestos awareness.

Asbestos fibres can cause cancer and lung diseases, such as asbestosis, so it is important to know where you may encounter items made of asbestos. There may be asbestos in:

- floor tiles
- artex or other decorative finishes
- asbestos cement pipes
- asbestos-containing fuse boards
- fire surrounds
- insulation storage compartments.

Figure 1.60 *Asbestos cement pipes*

If you suspect the presence of asbestos in the workplace then you should stop work immediately. An assessment should then be

Figure 1.61 *Asbestos-containing fuse-board*

Figure 1.62 *Asbestos in floor tiles*

made to decide who must do the work and you may need to engage the services of a licensed contractor. You will also need to minimize the spread of contamination to other areas and keep exposure as low as possible.

Remember

Asbestos fibres can cause lung diseases and cancer.

If you suspect asbestos is present then get it checked before you start work.

Only carry out work if you have been properly trained and have the right equipment.

Restoring the worksite on completion

Maintaining good relations with the customer is important right through to the completion of the contract and beyond.

Figure 1.63 *Restore the workplace to its original condition*

The work contract will usually have a completion date. In order to maintain good relations with the customer they should be advised if this is not going to be met.

When all the work is completed the worksite should be restored to an equivalent condition as when work started. This means that floor-boards must be replaced, walls replastered and walls and decorative finishes may need to be repainted.

If it has been necessary to breach fire barriers then they must be reinstated with a fire stopping material of an equivalent integrity.

Unused materials must be removed from the site and credited back to the stores.

Figure 1.64 *Fire barriers must be reinstated*

Figure 1.65 *Return unused materials to stores*

Waste materials

Waste materials must be disposed of in the most appropriate way. Many materials such as old conduit and switchgear may be recycled and therefore should not be sent to landfill sites.

Other waste such as fluorescent and other discharge lamps which contain highly toxic chemicals should be treated as hazardous 'non-special' waste. There are lamp-crushing and cleaning systems available which minimize many of the problems associated with the disposal of discharge lamps. Specialist licensed contractors may be required to dispose of hazardous waste materials.

Remember

Storing materials

A system should be implemented to protect materials and components from damage, during transit and storage or from deterioration, loss or incorrect identification.

● On arrival, items should be checked for damage and compliance with requirements and signed for on a delivery note.

● Components should be identified with job/customer and possibly drawing number.

● The storage area should be as clean and dry as possible, secure and with access only to authorized staff.

● There should be a procedure for the issue of stores to prevent loss, incorrect use and the mixing of stock.

● There should be frequent checks to detect any loss or degradation of stock.

Restoring the worksite

At the completion of the work the site should be restored to an equivalent condition, this could mean:

● replacing floorboards

● redecorating walls, skirting boards, etc.

● where fire barriers have been breached they must be replaced with a fire stop of the same integrity as the surrounding material

● removing unused materials and crediting them back into the stores

● disposing of waste materials in the most appropriate way.

Try this: Crossword

Across

1	Powerful water vapours (5)
4	We must do this to determine the risk (6)
6	Airborne particles which may cause lung problems (4)
8	Hazards may result in this being unacceptable (4)
9	This must be disposed of during site clearance (5)
10	These are used in electroplating baths (9)
13	The overall conditions in which we are working is this (11)
14	This may be required where the work area is airless (11)

Down

1	Where the work is carried out (4)
2	The air in the work location (10)
3	A common but dangerous material requiring special handling (8)
5	Where we keep the materials before installation (5)
7	The step we take to avoid danger arising (10)
11	This is generally stored on drums (5)
12	The place where it is positioned (8)

SELF ASSESSMENT

1   Powered mobile platforms may only be operated by persons over:

a. any age
b. 16 years
c. 18 years
d. 21 years

2   Cable on card drums should be stored:

a. on the ground
b. outside
c. raised off the floor
d. in the site office

3   Materials received on site should be signed for on:

a. a delivery note
b. a time sheet
c. an invoice
d. a requisition form

4   Fire barriers must be reinstated when a trunking run:

a. changes direction
b. is terminated
c. enters a distribution board
d. passes through a structural brick wall

5   Fluorescent lamps are best disposed of in:

a. a skip
b. a lamp crusher
c. the local bottle bank
d. a dustbin

Personal protective equipment and safety signs

RECAP

Before you start work on this chapter, complete the exercise below to ensure that you remember what you learned earlier.

A _____ is something that could potentially be harmful to a person's life, health, property or the environment.

List two examples of potential dangers on a work site:

_____

_____

How can materials and equipment be kept in a safe and good condition on site?

_____

_____

What does COSHH stand for?

_____

Asbestos fibres can cause _____ and _____ diseases.

LEARNING OBJECTIVES

On completion of this chapter you should be able to:

● Describe the procedures that should be taken to minimize or remove risks before requiring personal protective equipment (PPE).

● State the purpose of PPE.

● Specify appropriate protective clothing and equipment for identified work tasks.

● Describe safe practices and procedures using signs and guarding.

© Start Traffic Management

Figure 1.66 *Barriers*

Part 1 Legislation

The particular legislation we will be looking at in this chapter is:

● Personal Protective Equipment at Work Regulations (PPE)

● Noise at Work Regulations 1989

● The Health and Safety (Safety Signs and Signals) Regulations 1996.

Personal protective equipment

Personal protective equipment (PPE) is all the equipment intended to be worn or held by a person at work and which offers protection against one or more risks to their health or safety.

Wherever there are risks to health and safety that cannot be adequately controlled by any other means then PPE must be supplied and used.

This is the main requirement of the Personal Protective Equipment at Work Regulations.

Employers should consider all means of protection and assess whether PPE is required. It is only if there is no other way of providing the protection necessary to do the work safely that PPE should be supplied.

Employers are required to ensure that PPE is:

● properly assessed before use to ensure that it is suitable for the situation, fits the wearer and prevents or adequately controls the risks that may occur

● kept clean, stored and maintained in good repair

● provided with instructions for its correct use

● used in the proper manner by employees.

Figure 1.67 *Protective footwear*

© Pictures supplied by Draper Tools Limited

Remember

You have a duty to ensure that you know, understand and use the equipment provided correctly. If the equipment is provided it must have been assessed as necessary. Never fail to use the PPE provided because the job will only take a few minutes.

Footwear

Protective footwear should be worn where there is a risk of injury from manual or mechanical handling, electrical work or any work carried out in hot or cold weather conditions. When choosing footwear consideration should be given to:

● grip

● resistance to water or hazardous substances

● flexibility

● comfort for the wearer.

Protective footwear should be worn when there are risks due to:

● heavy materials being dropped on the feet

● penetration by nails and other sharp objects from below (standing on objects).

Eye protection

Safety goggles or spectacles are required in situations such as where there is:

● dust and flying debris, for example when chasing out brickwork

● sparks from cutting operations such as disc cutters

● chemical splashes, such as PVC conduit solvent

● airborne debris such as cement dust on construction projects.

Figure 1.68 *Eye protection*

© Pictures supplied by Draper Tools Limited

Remember

Eye protection may be required in dusty environments and in such conditions dust masks will also be required.

Hard hats

Head protection, more commonly referred to as hard hats, is required where there is a risk of injury from falling or flying objects such as:

● material being dropped by persons working above

- material being kicked into pits, excavations and service wells
- material falling from lifting activities such as hoists and cranes
- dangers from fixed protrusions in the work area, such as mounting brackets and pipe clamps.

Figure 1.69 *Head protection*

© Pictures supplied by Draper Tools Limited

Outdoor clothing

When working outside additional outdoor clothing may be required to:

- Offset the effects of wind and rain on the general health of the individual.
- Offset particular effects of the elements, long term exposure to the elements or prolonged outdoor working. These activities may involve work with little or no physical movement such as the termination of cables in a feeder pillar, or extreme physical activity such as digging.
- Protect against certain elements. This may also be provided by portable shelters, such as the framed work tent used by a cable jointer.

Figure 1.70 *Thermal socks*

Gloves

Leather gloves can safeguard against cuts resulting from manual handling of heavy and sharp objects.

Gloves may also be required when working outside in very cold conditions. Where there is a risk of an electric shock, electric arc or burns and the equipment cannot be made dead, then wearing suitable gloves may provide the protection required.

Hand protection is required:

- wherever activities may cause risk of skin penetration such as wood or steel splinters
- where the nature of the material may cause physical skin abrasion such as moving concrete blocks and bricks
- where the materials being handled may be hazardous by either being absorbed into the skin or by transfer to the mouth through contact with food.

Figure 1.71 *Hand protection*

High visibility clothing

High visibility clothing may be required:

- where work is carried out in an area with a high risk from passing traffic such as working on street furniture, street lights and bollards
- in areas with vehicular traffic such as loading bays and docks
- in areas where there are lifting operations being undertaken, such as the construction of high rise buildings.

Figure 1.72 *High visibility clothing*

Figure 1.73 *Some people may find that a woolly hat, in addition to a warm coat, is essential protective clothing!*

Ear defenders

Under the Noise at Work Regulations ear protection should be used in very noisy environments such as on a site where there are pneumatic drills, power presses or when using cartridge-operated tools and no other action to reduce the noise is possible.

A simple guide to determine when ear protection is needed is if:

● you have to shout to be heard by someone who is only 2 metres away

● the noise is intrusive for most of the working day or

● you work with or near sources of very loud noises such as pneumatic drills or cartridge-operated tools.

Figure 1.74 *Ear protection*

Ear defenders should comply with current BS specifications, fit properly and be comfortable.

Prolonged exposure to noise at work can cause temporary and possibly permanent hearing loss which may not be immediately obvious, but develops gradually over time. Another problem could be the development of tinnitus (ringing in the ears) which is a distressing condition.

Remember

When working in a noisy environment or wearing ear protection it will be harder for you to hear if people need to warn you of any dangers.

Ear protection is required where there is a high level of background noise, which may not be due to the activities you are involved in.

Try this: Crossword

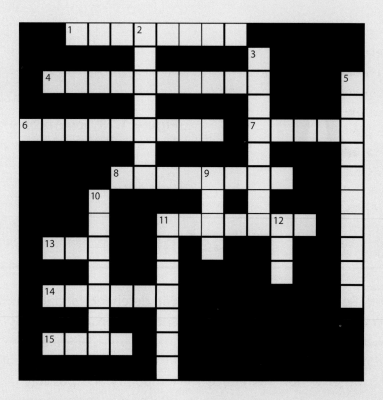

Across

1   Used to protect you from standing on sharp objects (8)

4   See 3 Down

6   See 3 Down

7   See 2 Down

8   See 10 Down

11   These can keep your eyes safe from dust (7)

13   See 15

14   These keep your hands safe (6)

15/13 Keep your head safe wearing this on a construction site (4,3)

Down

2/7   These will keep your toes toasty (7,5)

3/4/6 This is PPE (8,10,9)

5   See 9

9/5   Essential type of clothing when you are working adjacent to roadways (4,10)

10/8 When its wet and raining and you are working outside you need to wear this (7,8)

11   A pair of these will keep the flying bits out of your eyes (7)

12   One of a pair protected by both 11 across and 11 down (3)

Part 2 Safety signs

Safety signs will usually be found on construction sites where it is necessary to wear protective clothing. These are mandatory signs which are circular in shape with a white symbol on a blue background and they show what **must** be done. Other areas may not have signs and possible dangers must be assessed by the worker and appropriate protective gear worn.

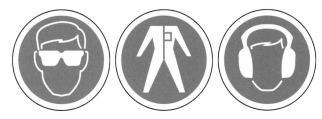

Figure 1.75 *Safety signs*

Table 1.3 shows the options available for protection for hazards that may be encountered when at work on site.

The Health and Safety (Safety Signs and Signals) Regulations 1996 state that all new Health and Safety Signs must contain pictorial symbols in addition to text. Other requirements contained in these regulations include the following:

● If a hazard cannot be adequately controlled by any other means then it must be marked with a safety sign.

● Every employer must ensure that their employees are trained in the meaning of safety signs.

● Any emergency escape route and fire fighting equipment must have their location identified with safety signs.

Table 1.3

	Type of hazard you may encounter	Options
Eyes	Chemical or metal splash, dust, projectiles, gas, vapour, radiation	Safety spectacles, goggles, faceshields, visors
Head	Impact from falling objects, risk of head bumping or hair entanglement	Helmets and bump caps
Breathing	Dust, vapour, gas, oxygen-deficient atmospheres,	Disposable filtering facepiece or respirator
Body protection	Temperature extremes, chemical or metal splash, impact or penetration, contaminated dust	Conventional or disposable overalls, boiler suits, specialist protective clothing
Hands	Abrasions, temperature extremes, cuts and punctures, impact, chemicals, electric shock, skin infection, disease or contamination	Gloves and gauntlets
Feet	Wet, slipping, cuts, punctures, falling objects	Safety boots and shoes with protective toe caps and penetration-resistant mid-sole

Task

List the PPE available at your place of work and where it is kept.

How is it stored?

_____

_____

_____

Where can you find instructions of how to use it correctly?

_____

_____

Prohibition signs

Prohibition signs have a crossbar through the centre and mean **stop** or **do not**. They are circular in shape, red on a white background and must be obeyed.

Figure 1.76 _No smoking_

Figure 1.77 _No pedestrians_

Figure 1.78 _Do not use_

Warning signs

Warning signs mean **caution**, **risk of danger** or **hazard ahead**. They are triangular and yellow with a black border.

Figure 1.79 _Danger – high voltage_

Figure 1.80 _Corrosive_

Figure 1.81 _Caution_

Task

Give examples of safety signs which are particularly appropriate to electrical installation sites.

_____

_____

_____

Mandatory signs

Mandatory signs mean **YOU MUST DO**.

They are circular in shape, with white text on a blue background.

Figure 1.82 _Wear eye protection_

Figure 1.83 _Wear head protection_

Figure 1.84 _Wear hand protection_

Safe condition signs

Safe condition signs are rectangular. They are white on a green background and indicate the correct action to be taken in an emergency or the correct place to go to. They give information about safe conditions.

Figure 1.85 _Emergency telephone_

Figure 1.86 _Indication of direction_

Figure 1.87 _First aid kit carried_

Task

Using suppliers information reproduce the current safety sign for each of the signs identified below.

A mandatory sign which means 'WEAR RESPIRATOR'.	An information (safe condition) sign which means 'EMERGENCY SHOWER'.
A prohibition sign which means 'DO NOT EXTINGUISH WITH WATER'.	A warning sign which means 'RISK OF EXPLOSION'.

You are likely to find extra information included with any of the safety signs, such as the type of first aid available, the particular type of eye protection required, or the clearance height of an obstacle.

Figure 1.88 *Emergency eye wash station*

Fire fighting equipment signs

The Health and Safety (Safety Signs and Signals) Regulations require the location of all fire fighting equipment to be marked in red.

Figure 1.89 *Fire extinguisher location sign*

Figure 1.90 *Fire alarm call point*

Figure 1.91 *Fireman's switch location sign*

Remember

You can help to keep yourself safe by:

- recognizing the hazards presented by the environment you are working in
- taking adequate precautions to protect yourself
- following any safety procedures set out for you by your employer
- taking notice of safety signs and obeying them.

SELF ASSESSMENT

1 The mandatory sign which means you must wear gloves shows:
   a. green gloves on a white background
   b. blue gloves on a white background
   c. white gloves on a green background
   d. white gloves on a blue background

2 Warning signs are:
   a. circular with a crossbar through the centre
   b. triangular with a black border
   c. rectangular with a green border
   d. circular with a blue border

3 Chemical hazards are covered by regulations called:
   a. COSHH
   b. BS7671
   c. EAWR
   d. HASAW

4 A prohibition sign:
   a. shows what must be done
   b. warns of danger
   c. gives information of safety provision
   d. shows what must not be done

5 The sign which indicates the position of an emergency telephone is:
   a. a white telephone on a blue background
   b. a white telephone on a yellow background
   c. a white telephone on a green background
   d. a white cross on a green background

1.6 Moving loads

Courtesy of Drapper Tools Ltd.

Figure 1.92 *Sack trolleys for manual handling*

LEARNING OBJECTIVES

On completion of this chapter you should be able to:

● Identify the method of moving a load related to its mass.

● Recognize the correct way to manually lift a load.

● Recognize the basic principal of the lever.

● Identify where pulleys should be used.

● Identify the safety requirements when using slings and pulleys.

● Recognize where barrows and trolleys can be used.

© Stephill Generators Ltd

Figure 1.93 *A diesel generator could be used to provide the power for a hoist*

Part 1 Manual lifting

What is a load?

A load is an object which has to be moved or lifted. One example of this could be where drums of 1.5mm^2 cable have to be moved from the floor onto a workbench.

Alternatively it could be a heavy electric motor which has to be moved from the stores into the workshop area.

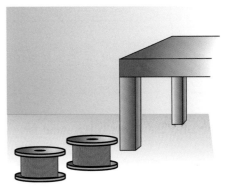

Figure 1.94 *Drums of cable to be lifted onto the bench*

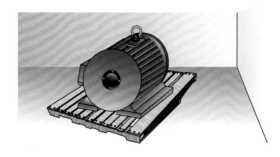

Figure 1.95 *Heavy motor to be moved to a new location*

Both of the above examples require a load to be moved from one place to another, but the methods used to achieve this would be very different.

Manually handling loads

It is important to recognize that moving a load involves a number of different considerations. The Manual Handling Operations Regulations 1992, as amended, recognize the possible risks involved in moving loads. By working through the flow charts and check lists shown in the Guidance on the Regulations it is possible to limit the risk of injury when carrying out manual handling operations. Many manual handling injuries arise from repeated handling using the wrong technique or incorrect posture. This can lead to permanent disability.

When considering moving a load the following question should be asked:

● Is it necessary to move the load?

If yes:

● assess the task and then
● reduce the risk of injury to the lowest reasonably practicable level, which may require the use of mechanical assistance.

It is generally accepted that loads over 20kg need lifting gear.

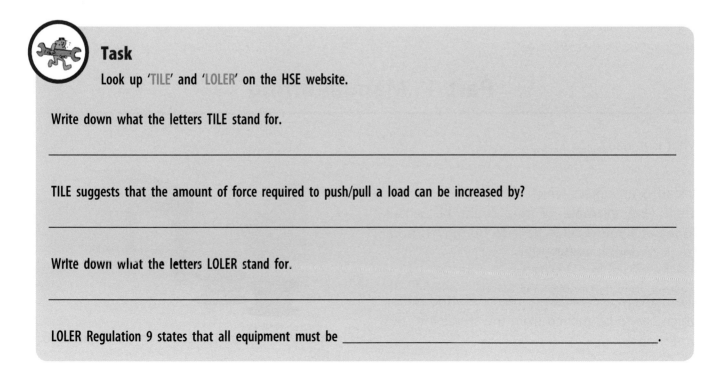

Task

Look up 'TILE' and 'LOLER' on the HSE website.

Write down what the letters TILE stand for.

_____

TILE suggests that the amount of force required to push/pull a load can be increased by?

_____

Write down what the letters LOLER stand for.

_____

LOLER Regulation 9 states that all equipment must be _____.

Remember

When moving loads:

● if it is appropriate to use lifting/transporting equipment and
● the employer has provided the equipment and
● you have been trained to use it

then you are legally required to use the equipment provided.

Figure 1.96 *An electrically powered hoist*

Figure 1.97 *Hydraulic jacks*

Note

Further guidance on lifting and handling is available from the HSE in 'Getting to grips with manual handling', INDG143 and 'Manual Handling, Manual Handling Operations Regulations 1992, (as amended), Guidance on Regulations, L23'.

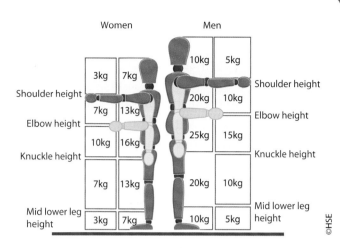

Figure 1.98 *HSE guidance on lifting and lowering weights from INDG143*

Guidelines

There are a number of factors that must be considered when a load has to be moved.

The task

Does the task involve moving the load in a way that may create a risk, for example does it mean that reaching up or a twisting movement Is involved? Are excessive distances involved? is there a risk of sudden movement of the load and will there be sufficient rest periods?

Figure 1.99 *Reaching up or twisting?*

The actual load

Could the load itself cause injury?

How can you find out how heavy it is? Is there any documentation with it or a label on it or can you estimate how heavy it is by looking at it?

Is it a bulky or difficult shape to get to grips with? Some objects may be greasy or wrapped in loose packing and these may need extra care. Has it got suitable handling points where you can get hold of the object? Has it any sharp corners or is it hot?

Figure 1.100 *Difficult to get to grips with*

Try this

Although loads must be risk assessed for the individual, the recommended maximum weight, close to the feet, that should be lifted by a woman from floor level is _____.

The recommended maximum weight, close to the feet, that should be lifted by a man from floor level is _____.

When lifting _____ has been provided to move loads you should be _____ in its use.

Part 2 The working environment

Is the route free from obstacles or what preparation work needs to be carried out to clear it?

Is the destination clear and does anything need to be arranged before starting off with the load?

Will there be enough headroom or will you have to stoop, and are there any fixtures or fittings in the way?

Is there any likelihood that the floor will be uneven or slippery?

Figure 1.101 *Keep the route free of obstacles*

Will you be carrying a bulky light load and could it be caught by a sudden gust of wind?

Consideration must also be given to factors such as the structure of the building, for example, uneven floor levels, the temperature and humidity and the level of illumination.

The capability of the staff involved

Are the operatives trained to carry out the task?

Consideration should also be given to the health and fitness of the operatives as this can affect the safety of the operation. Will the job place someone with a health problem at risk and does it require someone with unusual strength or height?

Other factors

Personal protective equipment or other additional clothing should only be used where it is absolutely necessary. For example, wearing gloves may impair dexterity although they may be required because of cold conditions or to avoid

injury from rough or sharp edges. Other protective clothing may make movement difficult.

Remember

Mass is the quantity of material in a body. The mass of the load is measured in kilograms (kg).

The weight of an object is the force which, because of gravity, it exerts on a platform placed under it.

The unit of force is the Newton (N).

Once an assessment of the task has been completed a decision must be made as to how the work is to be carried out safely. This may involve more than one person, the use of special equipment, preparation of the working area or the use of safety clothing and so on. Do not forget that if your task requires the use of appropriate equipment and this has been provided by your employer, and you have been trained to use it, then it is a legal requirement for you to do so.

Manual lifting

Whenever a load is to be lifted careful consideration needs to be given to the task. If it is to be lifted manually extra consideration needs to be given to the position of the body.

It is important to keep the spine in its naturally upright position. If the back is unnaturally arched forward there is a greater risk of injury to it. The knees should be bent so that when the legs are straightened, the load is lifted. In order to achieve this, the load should be kept close to the body with arms as straight as possible and both hands should be used.

Naturally upright

Figure 1.102 *Good lifting posture is vital*

To allow for the centre of gravity of the load the body should lean back slightly when lifting, allowing the body to act as a counter balance to the load.

When the load has been lifted and the body is straight care must be taken to avoid sudden turning or twisting as this can also cause damage to the back.

Remember

When lifting and carrying a load:

- STOP and THINK
- place your feet correctly and bend your knees
- keep your back in its naturally upright position
- keep your shoulders level and facing in the same direction as your hips
- keep your elbows in
- hold the load as close to you as possible
- position your hands so that your fingers do not become trapped
- use two hands and get a firm grip
- don't jerk
- put the load down before adjusting the position.

Carrying long loads

Although it would be good practice for two people to carry a long load, if it is being carried by a single person then care must be taken so that injuries do not occur. These injuries may be to the person carrying the long load or somebody else in the vicinity! The person carrying the load should ensure that the centre of gravity is directly related to the carrying position.

In Figure 1.103 the centre of gravity of the ladder is approximately above the shoulder on which it is being carried. To avoid unnecessary injury to others the front end of the load is kept high. However care must be taken when turning whilst carrying long items in this way.

Figure 1.103 *Good balance of a long load*

Lifting platforms

When a load has to be lifted for stacking or lifted onto the shoulders to be carried, it is useful to have a lifting platform so that the lift can be done in two stages. This will reduce the risk of strain or injury.

Figure 1.104 *Lifting platform*

Pushing and sliding

Lifting a load generally requires more effort than moving a load in a horizontal plane. When considering moving a load, look at all the possible alternatives, as in some cases it is easier to push or slide a load rather than lift it. If a load is to be pushed or pulled care must be taken not to damage the operative's back. To work at maximum efficiency with the least possibility of injury to the person, the back should be kept straight and the legs should do the pushing or pulling.

Figure 1.105 *Sometimes it is easier to push or slide a load*

Try this

There are a number of factors that must be considered when a load has to be moved, which include:

Is it _____ to move the load?

List **four** examples of risks that may be encountered with the load itself.

1 _____   3 _____

2 _____   4 _____

List **four** examples of risks that may be encountered on the route whilst moving a load.

1 _____   3 _____

2 _____   4 _____

Is the person who is to carry the load _____ of carrying it?

It is very important when manually lifting a _____ to consider the position of the body. Injury to the back can be caused by incorrect positioning, sudden turning or _____.

Lifting platforms can _____ the risk of strain or _____ by creating another _____ in the lifting process.

State **two** considerations which must be given to the position of the body when pushing a load.

1 _____

2 _____

Task

You are required to lift a bulky load of about 15kg from the floor and take it to a new position on a worktop 5m away. Assume the load is on the floor in front of where you are sitting now. Assess the situation, list any sources of hazard you feel you could encounter and describe how you would deal with them.

Part 4 Assisted moving

Pushing or pulling a load may be easier than lifting, but a heavy load on a flat surface can create a large friction resistance area.

This friction can be reduced if rollers are used between the load and the floor.

Levers

So that rollers can be placed under the load it has to be lifted one end at a time. The lifting can best be carried out with a lever. This is placed under one side of the load and then pushed down. As the part of the lever that is pushed down on is a great deal longer than the end that is lifting the load a mechanical advantage is created.

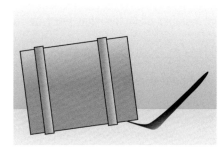

Figure 1.107 *Now a roller can go in place*

The lever may be of the bent type as shown in Figure 1.107 or a straight one used with a block as shown in Figure 1.108. The point at which the lever pivots is called the fulcrum.

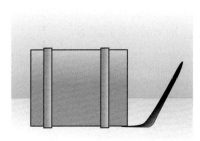

Figure 1.106 *Levers make lifting easier*

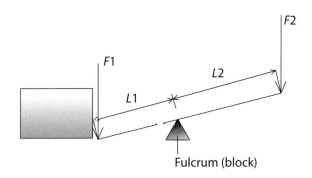

Figure 1.108 *Straight lever and fulcrum*

This method is rather like the principle of a see-saw, which is balanced when equal weight is placed on each end, i.e. it is in equilibrium. If a lever is the same length on either side of the fulcrum then a force equal to that of the load will create a state of equilibrium. When the force applied to the 'handle' of the lever exceeds that of the load, movement takes place and the load is lifted on one edge. To create a mechanical advantage, so that less effort needs to be used, the handle end of the lever is made longer.

A lever with a ratio of 4:1, that is to say, a lever as shown in Figure 1.108, where the handle (L2) is four times longer than the lifting arm (L1), needs only a quarter the effort to lift the load.

This can be calculated by the fact that to produce equilibrium:

$$F2 \times L2 = F1 \times L1$$

Consider a load of 10kg to be lifted by a lever, as shown in Figure 1.109.

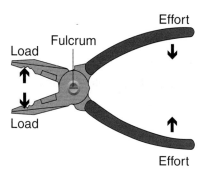

Figure 1.110 *Pliers*

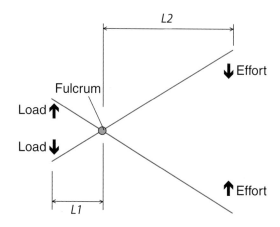

Figure 1.111 *Lever principle*

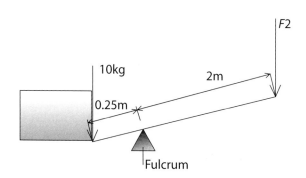

Figure 1.109 *Lever calculation*

As the handle of the lever is eight times as long as the blade the force required at F2 has been reduced by 8:1 and therefore the load can be raised using, 10kg ÷ 8 = 1.25kg.

Another example of how levers can be used is a pair of pliers or cutters, as shown in Figures 1.110 and 1.111.

Rollers

When the load has been raised rollers can be placed underneath as shown in Figure 1.112. The load can then be pushed gently forward on two or three rollers and further rollers placed under the front end as necessary.

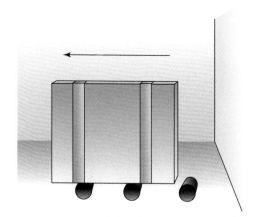

Figure 1.112 *Using rollers to move a load*

Try this

What is the load that can be lifted with an effort of 12kg on a lever with a distance of 0.3m between the load and the fulcrum and 0.6m between the effort and the fulcrum?

_____

_____

_____

_____

Remember

It is important to include all your calculations with your answer.

Barrows and trolleys

A wheelbarrow has a single front wheel or ball and can be used to carry heavy or bulky loads. It is also useful for carrying loose loads such as sand or gravel. Load the barrow so that it will not overbalance, use both hands, keep a straight back and steady the barrow before moving off.

Figure 1.113 _A wheelbarrow_

A sack barrow has two wheels and is more stable than a wheelbarrow. It is still very necessary to load the barrow correctly to avoid overbalance.

Figure 1.114 _A sack barrow or hand truck_

A flat trolley has four wheels and it is often used in stores where materials are constantly being moved. When on level ground a flat trolley is usually pulled whereas barrows are usually pushed. It may be necessary to prevent trolleys from moving at the wrong time in which case chocks (blocks or wedges) need to be placed to prevent the wheels turning.

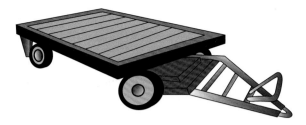

Figure 1.115 *A flat trolley*

Forklift truck

Another way of moving a load is by using a forklift truck. These are often used in large stores where the goods are to be stacked on pallets. Only authorized and trained personnel are allowed to use these.

Table trolley

A table trolley is a raised platform on four wheels, two fixed and two swivel braked to avoid the need for lifting the load. There will be a limit to the load that can be carried, which must not be exceeded – safe working load (SWL).

Courtesy of Sealey Power Products Ltd.

Figure 1.116 *A table trolley*

Slings and pulleys

When a heavy load has to be lifted vertically slings and pulleys can be used, but care must always be taken to ensure that the supports are capable of taking the maximum load. The maximum safe working load (SWL) should **NEVER** be exceeded.

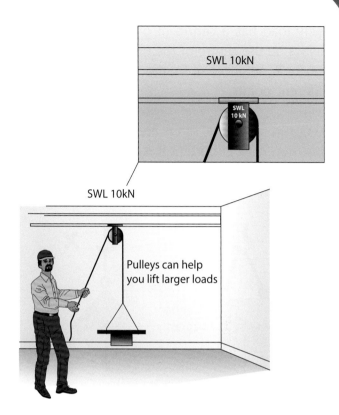

SWL 10kN

Figure 1.117 *Never exceed the SWL*

After the load has been lifted care must be taken to keep the area below the load clear and make sure that nobody can be trapped or crushed by the load.

Figure 1.118 *Keep clear!*

NEVER leave a suspended load unsupervised. Lower the load gently into position and, before you remove the lifting equipment, make sure that the load is stable and will not topple over.

Every load has a centre of gravity. This will not always be the centre of the object. The centre of gravity is determined by the shape and mass at different points.

To ensure even lifts, slings are used. Figure 1.119 illustrates how important it is to use slings sensibly as shown in (a), or the load will try to take up a new position as indicated in (b).

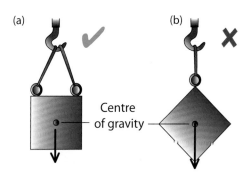

Figure 1.119 *The load at (b) would swing round and hang with the centre of gravity below the support*

Many objects that have to be lifted have sharp edges and corners. Rope slings should not be used unless they are protected as they may become damaged.

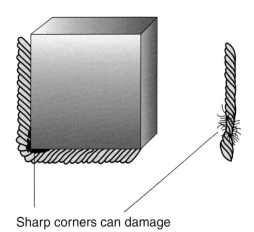

Sharp corners can damage

Figure 1.120 *Sharp edges can cause damage to ropes, slings and hands*

The pulley block

The pulley block consists of a continuous chain or rope passing over a number of pulley wheels as shown in Figure 1.121. Pulley blocks should be

regularly tested and their safe working load (SWL) displayed on them, which should **NEVER** be exceeded.

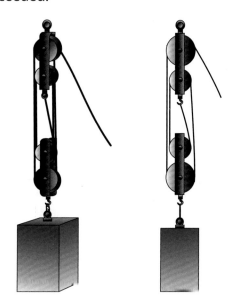

Figure 1.121 *Block and tackle lifting gear*

When loads are suspended on pulley systems they have a tendency to swing and twist. This problem is often overcome by having a stabilizing or control rope tied to the load and manned by a person given the sole responsibility of keeping the load straight.

Loads should never be left suspended in mid-air without someone to watch them. The area under the load should be kept clear at all times in case the load should fall.

When the load has been lowered gently into position it should be checked for stability before the sling is taken away.

Why use pulleys?

Pulleys can give several advantages when trying to lift an object. If a single wheel pulley is used there is no mechanical advantage but a difficult shaped object can be slung so that a single rope can lift it.

Where a pulley system with two pulleys is used the amount of effort required is only half that of a one pulley system. If we go to a four-pulley system like the one shown in Figure 1.121 the effort required is only a quarter (one fourth) of that required with a single pulley.

Example

When lifting a load of mass 16kg with a single pulley the effort required is also 16kg.

With two pulleys the same load can be lifted with half the effort:

i.e. 8kg

With four pulleys the same load can be lifted with one quarter of the effort:

i.e. 4kg.

Winches

A simple winch, as shown in Figure 1.122, consisting of a drum around which a rope is wound can also be used to raise loads. A crank handle rotates the drum and takes up or lets out the rope thus raising or lowering the load.

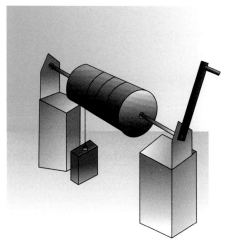

Figure 1.122 *Simple winch*

Power sources available

Lifting gear can be classified by its power source:

Mechanical: winches, pulleys, manual (muscle) power – probably yours!

Electrical: electric motors

Pneumatic: compressed air

Hydraulic: liquids

Petrol or diesel engines: although not used indoors because of the fumes.

 Try this

What mass can be lifted with an effort of 15kg on a two-pulley system?

_____

On a four-pulley system a load of 36kg has to be lifted. What effort is required?

_____

To lift a mass of 72kg an operative uses an effort of 18kg. What pulley system does he require to lift the load?

_____

Try this

Generally a person can carry loads of up to _____. Above that assistance is advisable. This may be extra people or _____ equipment.

List **three** examples of wheeled devices that are available for moving loads from one place to another.

1 _____

2 _____

3 _____

To lift loads up to a new height a lifting platform, _____ or _____ may be used.

Whenever loads are _____ care must be taken not to cause an accident or injury.

SELF ASSESSMENT

1   Which of the following is correct when manually lifting a load from the floor?

   a. straight legs and bent back
   b. upright back and straight legs
   c. bent legs and load at arms length
   d. bent legs and upright back

2   A load of 24kg is to be lifted using a lever and pivot as shown in Figure 1.123. The force that needs to be applied to the end of the lever to lift the load is:

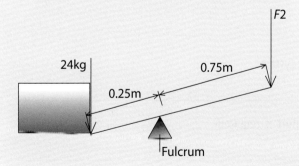

Figure 1.123

   a. 48kg
   b. 24kg
   c. 12kg
   d. 8kg

3   The effort required to lift a load of 40kg when using a four-pulley system is:

   a. 10kg
   b. 20kg
   c. 40kg
   d. 160kg

4   A table trolley is used to:

   a. lift a heavy load to position rollers
   b. shift a load to another location at a similar height
   c. transport a heavy load at low level
   d. move tall loads in the place of rollers or trucks

5   The generally accepted maximum load for a fit man to lift is:

   a. 5kg
   b. 15kg
   c. 20kg
   d. 50kg

Access equipment

1.7

RECAP

From what you learned in the previous chapter can you classify the following lifting devices by their power source?

Match each of the lifting devices to their power source:

Lifting devices: Block and tackle   Hydraulic lift   Winch

Diesel engine   Electric motor   Petrol engine

Power source	Lifting device
Manual power	_____
Electrical power	_____
Liquids	_____
Combustion	_____

LEARNING OBJECTIVES

On completion of this chapter you should be able to:

● Describe safe practices and procedures using access equipment.

● Describe situations and explain practices and procedures for addressing hazards when working at height.

● Identify suitable access equipment for specified situations.

● Describe the correct angle for erecting ladders.

● Describe methods of securing ladders.

● Recognize the need for safety requirements when working with tower scaffolds.

● Recognize the need for guardrails and toe boards on scaffolding.

Courtesy of Sealey Power Products Ltd.

Figure 1.124 *Access equipment*

Part 1

Access equipment provides a means of reaching the area where work has to be carried out.

The Working at Height Regulations 2005 apply whenever there is a risk of personal injury due to a fall. Work at 'height' applies everywhere a person can be injured from falling. Most people are aware that working at height regulations apply when working at any height **above** ground but it is important to note that they also apply in **all** instances where injuries from falling can occur – even **at or below** ground level.

Employers must do everything that is reasonably practical to prevent anyone falling. So they must carry out a risk assessment and plan and organize work at height so that:

- work at height is avoided where possible
- equipment or other methods are used to prevent falls where work at height cannot be avoided and the equipment is inspected and fit for use
- those that are to work at height are trained and competent

- account is taken of weather conditions that could endanger the health and safety of the operative
- any risk from fragile surfaces or falling objects are properly controlled.

A competent person is defined by the Construction (Health, Safety & Welfare) Regulations 1996 as follows:

Any person who carries out an activity shall possess such **Training, Technical Knowledge or Experience** as may be appropriate, or be supervised by such a person.

Employees are required to:

- report any safety hazard to their employer
- use the equipment supplied in a proper manner
- follow any training and instructions given.

One of the major contributory factors in accidents has been identified as a lack of training. It is important that those doing the work have knowledge and understand the importance of safe procedures.

Simple access equipment

Different access equipment will be required depending on the height at which work is to be carried out and the time the work will take. Reaching just above shoulder height may only require a step up, whereas scaffolding may be required for access to work on the roof of a building.

There are some basic rules which apply whichever access equipment is required.

- All access equipment should be set up on a firm level base.

- The equipment chosen must be suitable for the task so that the user does not have to overreach.

- All access equipment should be inspected regularly to ensure that it is in good condition. This does not just mean whether or not it is broken, it also includes checks such as looking to see if the surface is slippery because of mud or ice and other relevant hazards.

Remember

Whenever access equipment is to be used the manufacturer's instructions should always be followed.

Figure 1.126 *A pair of steps*

Step up

To reach up a short distance in comfort it is common to use a very simple piece of equipment called a step up, often referred to as a hop-up.

Courtesy of Sealey Power Products Ltd.

Figure 1.125 *Step up*

Figure 1.127 *Incorrect use of steps*

Steps

For work up to the height of a ceiling a pair of steps could be used. These should be high enough for the user to work without using the top three steps. Steps should be open to their fullest extent, locked where a locking or stay bar is provided and should be set on firm and level flooring.

HSE guidance on the use of steps includes:

- Use for short duration work (maximum 30 minutes) in one position.
- Light work no more than 10kg weight.
- Avoid working side on from steps.
- Do not overreach.

Trestles and platform

Quite often work at ceiling height involves working over some distance and in such cases trestles and a platform may be considered. They should only be used after a risk assessment shows that the risk of a person falling and injuring themselves is low.

If the risk assessment shows that the trestle system is suitable for the work then employers must also consider if the platform:

- is of sufficient dimensions
- is free from trip hazards or gaps through which a person or the materials being used could fall
- needs to be fitted with toeboards and guardrails. The height of the top guardrail should be at least 950mm and any unguarded gap should be no greater than 470mm.

If trestles and a platform are used they should be erected on firm ground, not loaded so they become deformed and the surface kept clean and tidy.

Trestle systems without guardrails would generally be unlikely to be used.

There is a requirement to provide a suitable means of access to the working platform. This could be by means of a separate pair of steps and this requirement is essential where the platform height is above 2m. These steps should exceed the height of the platform by at least 1m. An alternative, as shown in Figure 1.128, is a ladder fixed on a bracket made for a gate in the face of the platform.

The platform and trestles should always be dismantled before moving them to a new position.

Tested to BS EN13374-2004
and BS EN12811-1-2003

End Barrier Post

Reversible Gate and Post   S-Bracket

Langtons Ltd

Figure 1.128 *Platform and trestles with toeboards and guardrails*

Ladders

If access to somewhere higher is required then a ladder should be used. The parts of a ladder are shown in Figure 1.129.

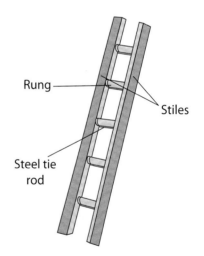

Rung

Stiles

Steel tie rod

Figure 1.129 *Ladder components*

It is very important when working from a ladder to remember the rule about standing it on firm level ground, and if it has to be put on soft earth then use a board. Never use boxes, drums or any other unsteady or sloping base.

Figure 1.130 *Using a board to stabilize a ladder on soft ground*

There are also a number of other key points to take note of:

● Ladders must not be too short as this could cause the user to overreach and fall off.

● Ladders should extend five rungs or 1.05m above the working platform unless there is an adequate handhold to reduce the risk of overbalancing.

Figure 1.131 *Do not overreach and maintain three points of contact on the ladder*

HSE guidance for the use of ladders includes the following recommendations:

● use only for short duration work in one position (maximum 30 minutes)
● always grip the ladder when climbing
● do not overreach
● light work, no more than 10kg
● do not work off the top three rungs.

When positioning a ladder the standard rule is the '1 in 4 rule', 1 unit out from the vertical for every 4 units up the vertical, which produces a ladder angle of 75°.

Ladders must be inspected frequently to ensure they are in good working order and that no rungs are damaged, missing or slippery. Damaged ladders should be withdrawn from use, and clearly labelled to show they are damaged and not to be used.

Aluminium ladders and steps are strong and lightweight alternatives but should not be used near electrical equipment or an electrical supply because of the possibility of creating a circuit and causing a short circuit or receiving an electrical shock. Fibreglass ladders and steps may be a suitable alternative for use in these locations.

Courtesy of Sealey Power Products Ltd.

Figure 1.132 *Fibreglass steps*

Try this

What would be the most appropriate access equipment to reach the ceiling of the room you are in?

_____

The most appropriate access equipment to reach a height of 5m for a short duration task would be?

_____

What would be the most appropriate access equipment to reach a height just above your own height?

_____

Suitable access to a trestle-supported platform would be by the use of?

_____

How far above the working platform without adequate handholds must a ladder extend?

_____

Part 2 Extension ladders and scaffolding

Extension ladders are often used providing convenience during transport and to reach higher levels where a single rise ladder would be impractical. Where extension ladders are used there must be an overlap of the rungs and this will depend on the length of the ladder.

For ladders with a closed length of up to 5m there should be two rungs overlapping as shown in Figure 1.133.

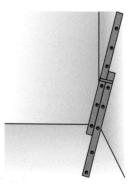

Figure 1.134 _Ladders with a closed length up to 6m_

For ladders with a closed length of up to 6m there should be three rungs overlapping as shown in Figure 1.134.

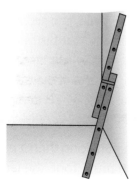

Figure 1.133 _Ladders with a closed length up to 5m._

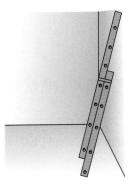

Figure 1.135 *Ladders with a closed length over 6m*

For ladders with a closed length of over 6m there should be four rungs overlapping as shown in Figure 1.135.

When moving ladders any distance they should be carried on the shoulders of two people, one at each end of the ladder. This will:

● distribute the weight of the ladder
● make it easier to control
● be less likely to cause injury to the persons carrying the ladder
● be less likely to cause danger, accident or injury when turning with the ladder.

Figure 1.136 *Two people are required to move long ladders*

Raising ladders

Ladders should be raised with the sections closed and two people may be required to raise the longer and heavier ladders.

One person stands on the bottom rung and holds the stiles to steady the ladder while the second one stands at the top and raises the ladder above head height and walks towards the bottom moving his or her hands down the ladder while walking.

Figure 1.137 *Correctly raising a ladder*

Remember
When the ladder is in position it must follow the 4:1 rule. This means that the height of the ladder above the ground must be four times the distance the ladder is out from the foot of the wall and the ladder will then be at an angle of 75° to the ground.

If the ladder is over 3m long then it must be secured or another person must 'foot' the ladder. To secure the ladder it must be lashed to a secure position like a scaffold pole as shown in Figure 1.139.

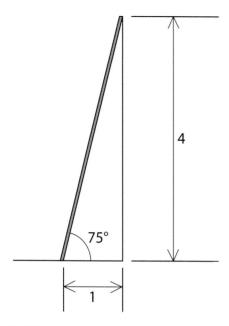

Figure 1.138 *Correctly erected ladder*

Figure 1.140 *Typical ladder stand-off*

Figure 1.139 *Ladder secured at the top*

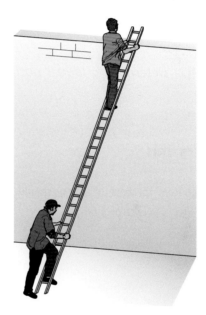

Figure 1.141 *Footing the ladder*

In some cases it is necessary to secure ladders of less than 3m, as even short falls can cause injuries.

A drainpipe or gutter is not a secure position and should not be used for the purpose of securing a ladder. In situations such as these ladder stand-offs can be used.

To foot the ladder another person must stand with one foot on the bottom rung, one foot on the ground and both hands holding the stiles as shown in Figure 1.141. The purpose of footing the ladder is to prevent it from moving, and the person carrying out this task must remain alert and observant at all times.

Example

An extending ladder is at an angle of 75° with the ground. How high up the wall will the ladder rest if it extends 1.5 m out from the base of the wall?

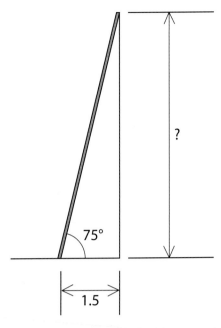

Figure 1.142 *Determining ladder position*

Answer

If the ladder is at the angle of 75° then the 4:1 rule applies. This means that the height of the ladder above the ground is 4 times 1.5m.

$$4 \times 1.5 = 6m$$

 Try this

How far out from the base of a wall should an extending ladder be positioned if the top end of the ladder is to rest on the wall at a height of 5.6m?

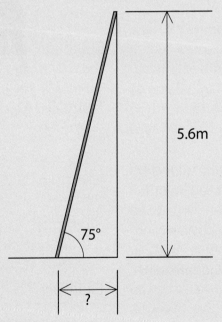

Figure 1.143 *Ladder position question*

Tower scaffold

For working at, and not just gaining access to, these higher levels, a tower scaffold can be used.

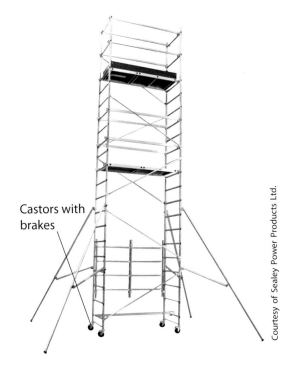

Castors with brakes

Courtesy of Sealey Power Products Ltd.

Figure 1.144 *Typical mobile access tower (commonly called a tower scaffold)*

Employers should provide training in the erection, use and inspection of the tower scaffold. PASMA, the Prefabricated Access Suppliers' and Manufacturers' Association, are the lead body and authority on matters relating to mobile access towers which are manufactured to BS EN 1004: 2004.

The assembly, dismantling or alteration of access towers should only be undertaken by a competent person. Operatives who have not been given the appropriate training should not erect or use the scaffold without suitable supervision.

The castors at the foot of the tower must be locked before anyone climbs the tower, and if someone should be working on the tower it must not be moved. When the tower has to be moved it should only be moved by pushing at the base.

Stabilizers or outriggers should be installed in accordance with the manufacturer's instructions. These will need to be lifted whilst the tower is moved but must always be reinstated and secured.

Any platforms used for working from must have toe-boards and guardrails for the safety of those working on and those below. Precautions must also be taken to ensure that access cannot be gained to incomplete scaffold towers and that only authorized persons can gain access at any time.

With this type of tower scaffold access to the working platforms must be made within the tower. Ladders or scaling the tower externally can result in the tower being overbalanced or falls and injury to the operative. Most tower scaffolds have internal ladders through the access platforms. The tower should only be climbed using the cross struts if the tower is specifically designed to allow this, otherwise the internal access ladders must be used.

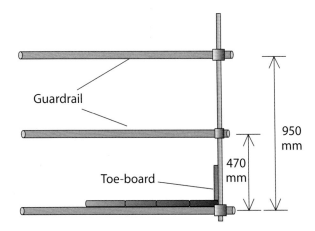

Guardrail

Toe-board

950 mm

470 mm

Figure 1.145 *Toe boards and hand rails must be fitted*

HSE guidance states the minimum height for handrails to be 950mm and that intermediate guard rails be fitted so that the unprotected gap does not exceed 470mm.

HSE recommends that an inspection of the tower should be carried out:

- after assembly in any positon
- after any event liable to affect its stability
- for towers under 2m to the working platform, at suitable intervals depending on the frequency and conditions of use
- for towers over 2m to the working platform, at intervals not exceeding seven days.

Mobile elevating work platforms (MEWPs) can provide a safe way of working at height. They allow the worker to reach the task quickly and easily, have guardrails and toe boards and can be used indoors or out. They include vehicle-mounted booms, cherry pickers and scissor lifts.

When working from platforms at height safety precautions need to be taken against injury in the event of a fall. For mobile elevating work platforms this generally takes the form of arrester harnesses which are secured to the platform supports. The safety harnesses must be designed and constructed to prevent serious injury in the event of a fall and allow the operative to work safely.

Figure 1.146 *Typical harness used on a MEWP at height*

Powered access platforms are a popular option for working at height in certain locations such as loading bays or tunnels. These should only be used by trained operatives.

Scaffolding

Full fixed scaffolding is another alternative when working at height but this should only be erected by a skilled scaffolder.

Note

Further information on the use of ladders and stepladders can be found in HSE Guidance available as free downloads from the HSE website www.hse.gov.uk

Try this

Choice of access equipment will depend on the use for which it is required.

Where the work requires simple access to a height we can use a step up, _____ or a _____.

For both vertical height and work on a horizontal plane we can use use either _____ and a platform or some form of _____.

The _____, dismantling or _____ of access towers should only be undertaken by a _____.

The erection of a ladder requires the application of the _____ rule to set it at the correct angle of _____ and the ladder should extend _____ rungs above the working platform.

All access equipment needs to be set up on a _____ _____ level base, suitable for the task and it must be in _____ condition.

Aluminium ladders should _____ be used near _____ or an electrical supply because of the possibility of causing a _____ circuit or receiving an _____.

Task

An electrician was working at a height which required the use of a ladder and he requested that his workmate foot the ladder. Whilst the electrician carried out his work his workmate was distracted and let go of the ladder. The ladder slipped and the electrician fell to the ground.

You were on site and witnessed the incident as it occurred and you have been asked to make a report to your company.

Write the report giving as many details as you can regarding the incident so that procedures may be put in place to ensure it does not happen again.

Your report should include such details as the type of work involved, the distraction and the extent of the injuries sustained by the electrician.

SELF ASSESSMENT

1   A ladder must extend past the working platform by a minimum number of rungs. The number of rungs is:

    a.   1

    b.   3

    c.   5

    d.   7

2   The maximum duration when working in one position on steps should be no longer than:

    a.   15 minutes

    b.   30 minutes

    c.   90 minutes

    d.   2 hours

3   When a ladder is leant against a wall the ratio of the height up the wall to the distance away from the wall at the bottom should not exceed a ratio of:

    a.   2:1

    b.   4:2

    c.   4:1

    d.   5:3

4   A mobile scaffold tower must be erected by:

    a.   anyone who can put it together

    b.   the site supervisor

    c.   the electrician who will use it

    d.   a competent person

5   When an extending ladder with a closed length of 3m is fully extended the minimum number of rungs overlap should be:

    a.   1

    b.   2

    c.   3

    d.   4

End test

1. **The Electricity at Work Regulations is a:**

   ☐ a. Statutory document

   ☐ b. Nonstatutory document

   ☐ c. Guidance material

   ☐ d. Best practice

2. **BS 7671, Requirements for Electrical Installations, is nonstatutory which means that it:**

   ☐ a. Will always result in prosecution if not complied with

   ☐ b. Cannot be cited in a court of law

   ☐ c. Is always legally binding

   ☐ d. Provides guidance

3. **An employer is required to report an accident to the Health and Safety Executive if an employee is:**

   ☐ a. Treated at work for a cut hand

   ☐ b. Sent home for 24 hours due to the accident

   ☐ c. Unable to work for more than three days due to the accident

   ☐ d. At a hospital casualty department for a morning before returning to work

4. **An Improvement Notice is issued by:**

   ☐ a. An employee to an employer

   ☐ b. An HSE inspector

   ☐ c. A works safety officer

   ☐ d. An employer to an employee

5. **An improvement notice has the effect of:**

   ☐ a. Stopping the activity identified on the notice

   ☐ b. Stopping all activity until the improvement is complete

   ☐ c. Setting a given time to achieve the improvement

   ☐ d. Advising that some improvement is required eventually

6. **The term COSHH regulations refers to:**

   ☐ a. Control of substances hazardous to health

   ☐ b. Centralizing of substances with health hazards

   ☐ c. Control of services, health and hazards

   ☐ d. Control of sustainable health and hygiene

7. **It is a legal requirement for an employer to produce a safety policy where they have a minimum of:**

   ☐ a. Two employees

   ☐ b. Three employees

   ☐ c. Four employees

   ☐ d. Five employees

8. **A power tool provided by an employer for an employee's use must be suitable under the:**

   ☐ a. Personal Protective Equipment at Work Regulations

   ☐ b. Provision and Use of Electrical equipment Regulations

c. Provision and Use of Work Equipment Regulations.

d. Personal Work Equipment Regulations

9. **The symbol shown means:**

a. No radios

b. No mobile phones

c. Site communications available

d. Risk of radiation

10. A white cross on a green background identifies the location of the:

a. Toilet facilities

b. Drying room

c. First aid box

d. Site office

11. The mandatory sign which means you must wear gloves shows:

a. Green gloves on a white background

b. Blue gloves on a white background

c. White gloves on a green background

d. White gloves on a blue background

12. Warning signs are:

a. Circular with a crossbar through the centre

b. Triangular with a black border

c. Rectangular with a green border

d. Circular with a blue border

13. A black panel on a fire extinguisher denotes it contains:

a. Water

b. Foam

c. CO2

d. Dry powder

14. An employee should report any accident or near miss in order to:

a. Ensure someone is made responsible

b. Ensure that the HSE is notified

c. Enable steps to be taken to prevent recurrence

d. Enable the person responsible to be punished

15. When carrying out the safe isolation of electrical equipment an electrician must comply with the requirements of the:

a. Working at Height Regulations

b. Construction Design and Management Regulations

c. Electricity at Work Regulations

d. Electricity Safety, Quality and Continuity Regulations

16. On discovering a hazard on site an apprentice electrician should:

a. Ignore it as its someone else's problem

b. Report it the next time they visit the office

c. Post a written report to the site office

d. Report it to a supervisor immediately

17. PPE equipment should be provided when:

a. A risk exists which cannot be removed

b. Removing risks would be time consuming

c. The risk is only for a short time

d. The operative requests it

18. When grinding metal an operative should always wear:

a. A hard hat

b. Eye protection

c. Safety shoes

d. Gloves

19. When wearing ear defenders one problem which may be encountered is not being able to hear:

☐ a. The radio

☐ b. The mobile phone

☐ c. Any danger warnings

☐ d. Any background noise

20. A suitable level of lighting of the work area is necessary to comply with the requirements of the:

☐ a. COSHH Regulations

☐ b. Workplace (Health and Safety and Welfare) Regulations

☐ c. Electricity at Work Regulations

☐ d. Manual Handling Operations Regulations

21. If you suspect the presence of asbestos in the workplace you should:

☐ a. Carry on and stay well clear of it

☐ b. Carry on but work carefully

☐ c. Remove it carefully and dispose of it

☐ d. Stop work immediately and report it

22. One of the risks associated with working with wet cement is:

☐ a. Weil's disease

☐ b. Asbestosis

☐ c. Dermatitis

☐ d. Myopia

23. The Electricity at Work Regulations permit working on or near live electrical equipment only when it is unreasonable in all cases for it to be dead and:

☐ a. Suitable procedures are followed

☐ b. It is only minor work

☐ c. The equipment is Class I

☐ d. Your supervisor say's it is OK

24. One of the dangers of working in an enclosed work space is the build-up of:

☐ a. Oxygen

☐ b. Nitrogen

☐ c. Carbon dioxide

☐ d. Hydrogen

25. A work mate has received an electric shock and is not breathing but is found to have a pulse. Your first action should be:

☐ a. Go and telephone for help

☐ b. Perform external cardiac compression

☐ c. Perform mouth to mouth and call for help

☐ d. Go and find the registered first aider

26. The system of supply shown above is:

☐ a. Extra low voltage

☐ b. Low voltage

☐ c. Reduced low voltage

☐ d. Nominal voltage

27. A method statement must contain the:

☐ a. Names of the operatives carrying out the work

☐ b. Date on which the work is to be carried out

☐ c. Details of the work to be carried out

☐ d. Name of the health and safety manager

28. A permit is issued under a permit to work system to ensure that employees carry out:

☐ a. Any work in the area identified in the permit

☐ b. No work other than that identified in the permit

☐ c. All work required in the area identified in the permit

☐ d. Work at any time in the area identified in the permit

29. When work is to be undertaken at height, which of the following is not an employer's responsibility?

☐ a. To perform a risk assessment before work is started

☐ b. To provide the right equipment for the job

☐ c. To provide suitable training for the task

☐ d. To provide employee's hand tools

30. It is generally accepted that moving loads over 20kg requires:

☐ a. Carrying handles

☐ b. A bent back

☐ c. Lifting gear

☐ d. Straight legs

31. A load of 24kg is to be lifted using a lever which has a length of 0.9m with the fulcrum set 0.3m from the load. The effort that must be exerted to lift the load is:

☐ a. 8kg

☐ b. 12kg

☐ c. 14kg

☐ d. 16kg

32. The term SWL on a hoist refers to the:

☐ a. Safe working load

☐ b. Standard working load

☐ c. Specified working load

☐ d. Suitable weight for lifting

33. When lifting a mass of 16kg using a block and tackle having eight pulleys the effort required will be:

☐ a. 16kg

☐ b. 8kg

☐ c. 4kg

☐ d. 2kg

34. HSE guidance on the use of steps recommends use in one position for short duration work of no more than:

☐ a. 15 minutes

☐ b. 20 minutes

☐ c. 30 minutes

☐ d. 40 minutes

35. A ladder set against a wall for fitting a lamp to a luminaire should be at an angle to the ground of:

☐ a. 45°

☐ b. 60°

☐ c. 75°

☐ d. 85°

36. PASMA are the lead body and authority on matters relating to:

☐ a. Mobile access towers

☐ b. Fixed scaffolding

☐ c. Extension ladders

☐ d. Trestle platforms

37. HSE guidance states that the handrails fitted to working platforms must be installed together with intermediate barriers so that the unprotected gap does not exceed:

☐ a. 950mm

☐ b. 790mm

☐ c. 630mm

☐ d. 470mm

38. The minimum age at which a person can operate a powered mobile platform is:

☐ a. 16

☐ b. 18

☐ c. 20

☐ d. 25

39. A mobile tower over 2m in height should be inspected at intervals no longer than:

☐ a. 3 days

☐ b. 5 days

☐ c. 7 days

☐ d. 10 days

40. Aluminium ladders should not be used to work near electrical equipment due to the risk of:

☐ a. Falls

☐ b. Electric shock

☐ c. Overreaching

☐ d. Insecure support

Unit two

The environment

Material contained in this unit covers the knowledge required for C&G Unit no. 2351-302 (ELTK 02), and the equivalent EAL unit.

This unit considers the environmental legislation, our impact upon the environment from our work activities and the steps we can take to make our work processes, the installation and the use of resources more environmentally friendly.

You could find it useful to look in a library or online for copies of the legislation and guidance material mentioned in this unit. Read the appropriate sections and remember to be on the lookout for any amendments or updates to them.

Before you undertake this unit read through the study guide on page x. If you follow the guide it will enable you to gain the maximum benefit from the material contained in this unit.

2.1

Environmental legislation, practices and principles

LEARNING OBJECTIVES

On completion of this chapter you should be able to:

● Specify the current relevant legislation for processing waste.

● Describe what is meant by the term environment.

● Describe the ways in which the environment may be affected by work activities.

Part 1

In general terms the 'environment' relates to the surroundings of an object or person. There is an increasing need for people to consider the effects of their actions on the environment of the planet. This means considering the impact of our actions, materials, rubbish and waste and bearing in mind how this affects the environment.

Pollution of the environment

The Environmental Protection Act gives the following definition of pollution of the environment:

> 'Pollution of the environment' means pollution of the environment due to the release (into any environmental medium) from any process of substances which are capable of causing harm to man or any other living organisms supported by the environment.

In order to control our impact on our environment there is a considerable array of legislation relating to environmental protection. Much of this legislation may, at first glance, appear to be unrelated to the electrical installation industry. However the design and construction of an electrical installation must take account of this legislation as must the maintenance and repair of electrical installations. This legislation considers the materials used, the disposal of waste and any redundant materials and equipment.

As with all legislation the Acts and Regulations are subject to review and updating so it is important to make sure we refer to the latest edition when establishing any requirements. We need to be aware of the applicable legislation and we shall consider those most relevant here.

Remember

Guidance material is also subject to change as legislation, technology and materials change. Make sure the guidance you refer to is the current version.

Environmental Protection Act 1990

The Environmental Protection Act 1990 aims to control and reduce pollution. It is an act of parliament which sets the fundamental requirements for the structure and authority for both waste management and the control of emissions into the environment.

Part I of the Act establishes the regime in which the Secretary of State for the Environment, Food and Rural Affairs sets limits regarding emissions into the environment. Authorization and enforcement is the responsibility of the Environment Agency (EA) and the Scottish Environment Protection Agency (SEPA).

Operation of any prescribed process without approval is prohibited and there are criminal sanctions against offenders.

Figure 2.1 *Air pollution in action*

Part II of the Act concerns the regulation and licensing of the disposal of controlled waste on land. Unauthorized or harmful depositing, treatment or disposal is prohibited and enforceable by criminal sanctions.

Task

Contact your local authority and find out what their policy is on the disposal of fluorescent and low energy lamps and record this requirement.

_____

_____

There is a duty of care on the importers, producers, carriers, keepers, processors or disposers of controlled waste to prevent unauthorized or harmful activities. A breach of this duty of care is a crime.

Local authorities have duties to collect controlled waste and to undertake recycling. There are criminal penalties on households and businesses who fail to cooperate with local authorities' arrangements.

The Hazardous Waste Regulations

Status: *This is the original version (as it was originally made). UK Statutory Instruments are not carried in their revised form on this site.*

STATUTORY INSTRUMENTS

2005 No. 894

ENVIRONMENTAL PROTECTION, ENGLAND AND WALES

The Hazardous Waste (England and Wales) Regulations 2005

Made - - -	*23rd March 2005*	
Laid before Parliament	*24th March 2005*	
Coming into force in accordance with regulation 1(1)		

The Secretary of State, being a Minister designated(**1**) for the purposes of section 2(2) of the European Communities Act 1972(**2**) in relation to measures relating to the prevention, reduction and elimination of pollution caused by waste, in exercise of the powers conferred on her by section 2(2) of that Act and section 156 of the Environmental Protection Act 1990(**3**), makes the following Regulations:

Source: www.legislation.gov.uk

Figure 2.2 *The Hazardous Waste Regulations*

A hazardous waste is defined as waste that poses substantial or potential threats to public health or the environment.

There are four factors that determine whether or not a substance is hazardous:

- Is it flammable?

- Is it corrosive?

- Is it toxic?

- Is it reactive?

© HSE

The Hazardous Waste Regulations 2005 (amended 2009) identify the requirements for dealing with waste classified as hazardous, and these are identified in Annexes I, II & III of these regulations.

Pollution Prevention and Control Act (PPCA)

Part I of this act is concerned with regulating pollution from industrial processes and it considers:

- preventing or minimizing pollution of the environment due to the release of substances into the air, water or land

- preventing or minimizing air pollution (this applies to those industrial processes which are not considered to give rise to significant pollution of water or land).

Part II of the act contains the Waste Management Licensing system, which is concerned with regulating the deposit, disposal or recovery of waste.

In addition to the PPCA there is the Planning Policy Statement 23: Planning and Pollution Control (PPG23). This document, produced by the British Government, is intended to complement the pollution control framework under the Pollution Prevention and Control Act and the Pollution Prevention and Control Regulations.

Control of Pollution Act

This act primarily covers the legislation relating to waste on land, pollution of water and noise pollution.

Control of Noise at Work Regulations 2005

These regulations concern the levels of exposure to noise in the workplace, setting maximum levels, limitation and protection with the intention of minimizing the detrimental effects on operatives' hearing.

Packaging (Essential Requirements) (Amendment) Regulations 2009

These regulations identify the requirements for packaging material in respect to their content and

suitability for safe disposal and minimizing the impact of the materials used on the environment.

The Environment Act

The Environment Act provides for the establishment of:

- The Environment Agency
- The Scottish Environment Protection Agency.

It also makes provision in respect to:

- contaminated land and abandoned mines
- National Parks
- the control of pollution
- the conservation of natural resources
- the conservation or enhancement of the environment.

The Waste Electrical and Electronic Equipment Directive (WEEE Directive)

Not for disposal in landfill

This is the European Community Directive 2002/96/EC on Waste Electrical and Electronic Equipment (WEEE) which, together with the Restriction of Hazardous Substances (RoHS) Directive 2002/95/EC, became European Law in February 2003, setting collection, recycling and recovery targets for all types of electrical goods.

The Waste Electrical and Electronic Equipment (WEEE) Regulations apply to electrical and electronic equipment with a voltage of up to 1000V ac or 1500V dc.

You will need to comply with the WEEE regulations if you produce, handle or dispose of waste that falls under one of ten general categories of WEEE:

- small household appliances
- large household appliances
- electrical and electronic tools
- consumer equipment
- lighting equipment
- monitoring and control equipment
- IT and telecommunications equipment
- toys, leisure and sports equipment
- medical devices
- automatic dispensers.

Examples of the products which fall within these general categories are given in Schedule II of the WEEE Regulations.

Figure 2.3   *Disposal under WEEE will be required*

Our interaction with the environment

Having examined some of the relevant legislation we also need to consider how our work activities can affect the environment. To do this we shall look at three main areas where we interact with the environment: land, air and water courses.

Land contamination

Whilst recycling continues to reduce the quantities, much of the waste we produce is disposed of in landfill sites, or rubbish tips as they are commonly called. To comply with the statutory requirements these landfill sites have to be registered and the material placed in them controlled. The disposal of certain materials results in harmful substances leeching into the soil. This may then render it unsafe for people and livestock and often affects the plants that are able to grow there. To ensure that this does not happen, the disposal of the 'hazardous waste' is dealt with separately with much of it being recyclable.

Even simple actions such as throwing disposable batteries, oil and lead-based paints, used motor oil or discharge lamps in with the landfill waste can seriously damage the environment. In the electrical industry there are a number of such materials in everyday use and care must be taken to ensure the hazardous materials are disposed of by authorized contractors using methods in accordance with the legislation.

Figure 2.4 *Typical landfill*

Air pollution

Air pollution takes into account the effect of air-borne waste on the quality of the air we breathe and its effect on the atmosphere. Air quality has greatly improved over the years starting with the introduction of smokeless fossil fuels and subsequently due to more use of gas and other alternatives. Much of the old-style burning of rubbish has ceased and with it an improvement in air quality. Burning the PVC insulation from cables and conductors, whether intentionally or as the result of a building fire, has serious implications.

Local authorities should be consulted regarding their policy on bonfires, this may range from simply regulating what material may be burnt or the time at which bonfires may be lit through to a total ban on bonfires. It is important to make sure the regional policies are complied with and, if possible, avoid burning rubbish altogether.

PVC plastic also requires toxic additives to make it stable and usable. These additives are released during the use (and disposal) of PVC products, resulting in increased human exposures to phthalates, lead, cadmium, tin and other toxic chemicals.

There are other 'hidden' emissions into the atmosphere in the form of gases which are not visible to the naked eye. The effect of the release of propellant gases into the atmosphere has been well documented and their use reduced. The release of gases such as Freon into the atmosphere is now controlled by legislation as 'controlled substances'.

Domestic air pollution has reduced considerably but industrial processes still produce pollutants. Many of these are due to production processes; the generation of electricity by coal fired power stations, iron and steel works and chemical processes are examples. Legislation is set to bring the levels down still further and continue reducing air pollution.

Water courses

Much of the material disposed of in drains makes its way inevitably into the water system and many materials contain harmful chemical components which are not removed by standard filtering methods. The disposal of paints, oils, caustic substances and the like down drains should be avoided at all times. These materials must be disposed of by an approved method.

Task

Contact your local authority and find out what their policy is on bonfires and record this requirement.

_____

_____

Similarly, materials disposed of into or onto the ground are eventually washed into the waterways by rain. The same criteria apply to the disposal of these hazardous materials: they should be removed by an approved method and not just tipped onto or buried into the ground.

It is also inevitable that some airborne contaminants will be brought to ground with rain and therefore make their way into the water system.

Figure 2.5 *Water courses can easily be polluted*

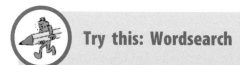

Try this: Wordsearch

Find the words from the list below in the wordsearch puzzle:

```
A I R P O L L U T I O N U D D
F E I V A C O R R O S I V E T
L V K G H Y H Y T J E W T K B
A E R S I R P I Z S S U R L W
M W R L P O R D R U L V L Z U
M K U N N E M U Y L U I D U A
A Q Z B E J O V O B F G N A T
B E O E D C R P G D L M C S M
L J W Q R T X J N T O X I C O
E O Y E R Y A A M L Z K I E S
E Q T S R Y L E Z K T H H Z P
Z A L E G I S L A T I O N P H
W E N V I R O N M E N T A D E
H A Z A R D O U S W A S T E R
A I B X G T Y X Y Y E J F I E
```

AIR POLLUTION	ENVIRONMENT	LANDFILL	TOXIC
ATMOSPHERE	FLAMMABLE	LEGISLATION	WATER COURSE
CORROSIVE	HAZARDOUS WASTE	POLLUTE	WEEE

Correctly complete the questions below before going on to the next chapter.

SELF ASSESSMENT

Circle the correct answers.

1   The Act which defines the limits for emissions into the environment is the:

a.   Health and Safety at Work etc. Act

b.   Environmental Protection Act

c.   The Buildings Act

d.   Prevention of Pollution Act

2   Information relating to the disposal of controlled waste in your location may be obtained from the:

a.   Environmental protection act

b.   WEEE directive

c.   Local authority

d.   Environmental health office

3   One of the factors which determines whether a product is classed as Hazardous Waste is if it is:

a.   inert

b.   powdery

c.   corrosive

d.   non-combustible

4   The WEEE directive applies to equipment operating at voltages up to and including:

a.   230V ac

b.   100V ac

c.   500V ac

d.   1000V ac

5   The disposal of transformer oil in with general landfill waste may result in the contamination of the:

a.   land and atmosphere

b.   atmosphere and watercourses

c.   land and water courses

d.   air and land

2.2

Building regulations

RECAP

Before you start work on this chapter, complete the exercise below to ensure that you remember what you learned earlier.

The Act which is concerned with preventing or minimizing pollution of the environment is the?

_____

The disposal of electrical equipment operating at 230V ac must comply with which directive?

_____

The three main areas of pollution of the environment are?

_____

LEARNING OBJECTIVES

On completion of this chapter you should be able to:

● Identify the requirements of the Building Regulations in respect to electrical installations.

● Identify the requirements of the Code of Practice for Sustainable Homes in respect to electrical installations.

Part 1   The Building Regulations

The Building Regulations are made under the Building Act 1984 and apply in England and Wales. Their purpose is to:

- secure the health, safety, welfare and convenience of people in or about buildings or others who may be affected by buildings or matters connected with buildings
- further the conservation of fuel and power
- prevent waste, undue consumption, misuse or contamination of water.

The Building Regulations apply to the design and construction of buildings and the provision of services or fittings in connection with buildings. They do not apply to existing buildings unless these are added to or altered, and then only in relation to making sure the new work being carried out is compliant.

In addition to the statutory document, the building regulations contain a number of Approved Documents relating to building work. Those Approved Documents which generally relate to the electrical installer are:

- A: Structure
- B: Fire safety
- E: Resistance to the passage of sound
- F: Ventilation
- L1: Conservation of fuel and power in dwellings
- M: Access to and use of buildings
- P: Electrical safety.

Not all of the content of these Approved Documents applies to the electrical installer and some does not relate directly to the environment or sustainability. However, we shall consider the areas relevant to the electrical installer in brief in this chapter.

Structure

This Approved Document provides information on the requirements in connection with the building structure. It achieves this by referring to the requirements of a number of British Standards including BS 5268, which has now been superseded by BS EN 1995. The key areas for electrical installation are the drilling and notching of joists and the chasing of walls.

Approved Document A sets out these requirements which are summarized in Figures 2.6–2.8 below.

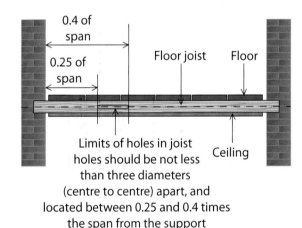

Figure 2.6 *Limitations for holes in joists*

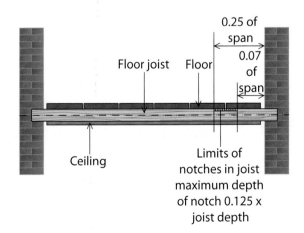

Figure 2.7 *Limitations for notches in joists*

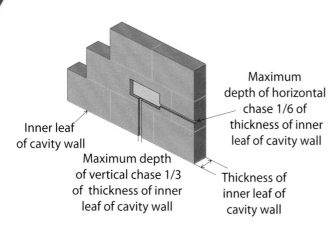

Inner leaf
of cavity wall

Maximum
depth of horizontal
chase 1/6 of
thickness of inner
leaf of cavity wall

Maximum depth
of vertical chase 1/3
of thickness of inner
leaf of cavity wall

Thickness of
inner leaf of
cavity wall

Figure 2.8 *Chases in internal walls*

Fire safety

Approved Document B is in two volumes with Volume 1 considering the requirements for 'dwellinghouses' and Volume 2 relating to buildings other than 'dwellinghouses'.

The general requirements are for buildings to have:

- means of warning and escape
- a structure that retains its stability for a reasonable period in the event of fire
- internal linings which inhibit the spread of fire within the building
- a roof and external walls that adequately resist the spread of fire
- reasonable facilities to assist fire fighters in the protection of life
- reasonable access provision for fire appliances to gain access to the building.

Volumes 1 and 2 include information relating to the fire detection and alarm systems and this is contained in the following sections:

- B1 Means of warning and escape
- B2 Internal fire spread (linings)
- B3 Internal fire spread (structure)
- B4 External fire spread

- B5 Access and facilities for the Fire and Rescue Service.

As stated above the main difference in the content is that Volume 1 relates to the requirements for dwellinghouses and Volume 2 is for all other buildings. There are some similarities in both volumes, but it is true to say that the requirements from Volume 2 are more onerous than those from Volume 1. However, irrespective of their type, more complex installations require greater zoning, defined escape routes and so on.

> **Note**
>
> The technical requirements for the installation of fire systems are given in BS 5839.

In dwellings there is a requirement for the electrical installer to ensure that, where appropriate, a suitable fire alarm system is installed. This may be a simple system with linked smoke detectors or a more complex, multi-zoned installation including a combination of smoke, heat and rate of rise detectors.

Where cables are installed through fire barriers the installer should ensure that the fire integrity of the fire barrier is maintained. These fire barriers include internal walls and ceilings which provide fire separation. So, for example, the ground floor ceiling of a two-story dwelling is a fire barrier between the floors. Where cables pass through the ceiling the integrity of the fire barrier must be maintained. Downlighters recessed into the ceiling should either:

- include integral fire protection (recommended) or
- be provided with a suitable intumescent hood which allows suitable ventilation and provides fire protection.

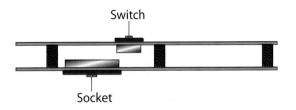

Figure 2.10 *Sharing the same cavity*

Figure 2.9 *Typical intumescent fire hood*

Figure 2.11 *Typical internal fire protection pad*

A similar requirement exists where accessories are recessed into partition walls which are load bearing or form a 30-minute fire barrier. This is particularly significant where accessories are installed back to back in the same cavity of the wall. Where accessories are installed in this way, as shown in Figure 2.10, they should include either a back box which incorporates fire protection or have a suitable fire protection pad installed as shown in Figure 2.11.

Remember

The depth of the box must be selected so as to not cause strain or undue pressure in the cables due to the addition of the fire protection pad.

Try this

Circle the correct answers for multiple choice questions 3, 4 and 5.

1 Identify the Approved Documents which relate to:

   a. Fire safety   _____

   b. Ventilation   _____

   c. Electrical safety   _____

2 List the **two** common methods used to provide fire protection where accessories are inserted in walls forming a fire barrier.

   _____

   _____

3   The maximum depth for a horizontal chase in the internal leaf of a cavity wall is based upon the thickness of the block. For a 90mm block the maximum depth is:
   a. 7.5mm
   b. 15mm
   c. 30mm
   d. 60mm

4   The location of a hole drilled in a joist with a span of 5 metres must be at a distance of no less than what from the supporting wall?
   a. 0.75m
   b. 1.00m
   c. 1.50m
   d. 1.75m

5   The section of the Approved Document relating to the internal spread of fire through the linings of the building fabric is B:
   a. 1
   b. 2
   c. 3
   d. 4

Task

Look around your home or workplace and identify three areas where there may be an internal fire barrier between rooms and record these below.

1 _____

2 _____

3 _____

Part 2

Resistance to the passage of sound

Having looked at the requirements to minimize the spread of fire, the next area for the electrical installer to consider is the passage of sound. In much the same way as installing cables and accessories can encourage the spread of fire they may also allow sound to travel through the building.

The requirements of Approved Document E affect the electrical installer in two ways:

1. The inclusion of sound insulation within the building structure acts in a similar way to thermal insulation as far as the cables are concerned. The cable is unable to give off as much heat as it would if installed in free air

and as a result the current carrying capacity of the cable may need to be reduced. This must be considered at the design stage of the installation.

2. Any accessories installed must maintain the sound insulation integrity, much the same as is required for the fire integrity. So recessed lights, for example, must not reduce the sound insulation.

The requirements of Approved Document E may be met when installing suitable fire barrier reinstatement for the electrical installation. The installer should ensure that the fire stopping material also meets the requirements for sound penetration. If it fails to do so additional measures will need to be taken. However, consideration must be given to the requirements of Approved Document E for all installations including those where no fire barriers have been breached.

Ventilation

Ventilation is the next area to consider. The advent of double-glazed windows which seal against the elements and provide better insulation, means the need for ventilation has increased. No longer does air circulate in and out of a building through ill fitting doors and windows and so a structured form of ventilation is required.

Controlled ventilation may also provide the opportunity for heat recovery and so improving the property's carbon footprint and sustainability.

Approved Document F requires adequate ventilation to be provided for people inside buildings. It does not apply to building spaces:

- which are used solely for storage
- where people do not normally go

- forming a garage used in connection with a single dwelling.

The Approved Document identifies the requirements related to ventilation in two sections: Section 1, Dwellings and Section 2, Buildings other than dwellings. Section 3 considers work on existing buildings such as adding rooms. The main topics being as follows:

Dwellings:
- ventilation rates
- ventilation systems for dwellings without basements
- ventilation systems for basements
- ventilation of habitable rooms through another room or a conservatory.

Buildings other than dwellings:
- access for maintenance
- commissioning
- offices
- ventilation rates
- natural ventilation of rooms
- mechanical ventilation of rooms
- alternative approaches
- ventilation of other types of buildings
- ventilation of car parks
- alternative approaches for ventilation of car parks.

From the electrical installer's viewpoint many of the ventilation requirements will involve the provision of electrically operated and controlled ventilation systems. Figure 2.12 shows the most basic system for a two-storey dwelling which comprises background ventilators and intermittent extract fans.

A fan in a bathroom is installed to provide extraction and will often require a run-on facility to ensure adequate air change, see Figure 2.13.

Figure 2.12 *A basic ventilation system*

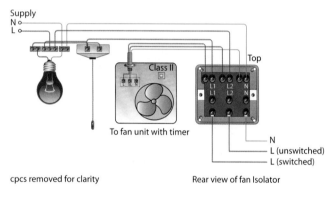

cpcs removed for clarity   Rear view of fan Isolator

Figure 2.13 *Typical bathroom fan control*

Any ducting must be correctly installed and supported to ensure there are no restrictions to the flow and there are no irregularities in the run where moisture can accumulate, see Figures 2.14 and 2.15.

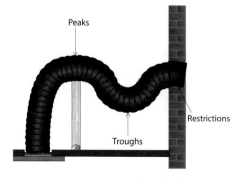

Figure 2.14 *Incorrect method of ducting*

Ducting should be as straight and as tight as possible to reduce resistance and maintain fan performance.

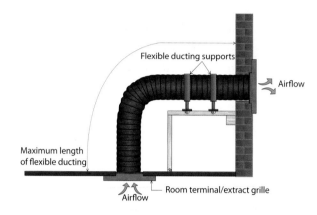

Figure 2.15 *Correct method of ducting*

Conservation of fuel and power in dwellings

The conservation of fuel and power is the subject of Approved Document L which comprises:

● L1A: New dwellings

● L1B: Existing dwellings

● L2A: New buildings (other than dwellings)

● L2B: Existing buildings (other than dwellings).

All of which provide guidance on meeting the requirements or conservation of fuel and power.

These requirements are related to carbon dioxide (CO_2) emissions, and the type and quantity of installed current using electrical equipment will affect the CO_2 emissions. The Approved Document identifies the target carbon dioxide emission rate (TER) for the building and the building emission rate (BER) or dwelling emission rate (DER) should not be greater than the TER.

Remember that the CO_2 emissions also include those produced in generating the electrical supply to the property. Reducing the energy used will reduce the emissions.

If we consider the requirements for new dwellings, one of the key considerations relates to the fixed lighting, both inside and outside. The recommendation is that internal lighting points should only accept lamps having a luminous efficacy greater than 40 lumens per watt, such as compact fluorescent lamps.

For a new dwelling a good approximation is that one in every four lamps should be energy efficient lamps.

For external lighting the requirement is that the lamp capacity should not exceed 150W and be automatically switched off when there is sufficient daylight and when it is not required at night (passive infra red detector control for example). Alternatively the luminaires should only accept lamps having an efficacy greater than 40 lumens per watt.

This requirement also applies to existing installations when they are extended or when an existing lighting circuit is replaced. It would also apply where a new dwelling is created due to a change of use.

The phasing out of the poor efficacy tungsten filament lamps is intended to reduce the availability of the poorer efficacy lamps by removing them from the marketplace.

(a)

(b)

Figure 2.16 *(a) Typical low energy lamps. (b) Typical GU10 LED lamp*

The replacement of existing tungsten halogen lamps with LED lamps will contribute considerably to the saving in terms of energy consumption and lamp costs over time.

Try this: Crossword

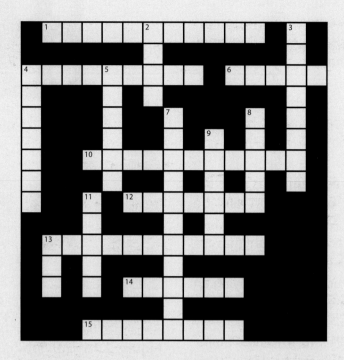

Across

1   Where these are flush mounted we must consider the spread of fire (11)

4   We are trying to reduce these to protect the environment (9)

6   We must not cut this too deep in the wall (5)

10   When we make one of these through a fire barrier we must reinstate afterwards (11)

12/13 This type of lamp is a good energy saver (7,11)

14   If making one of these through an acoustic barrier it will need to be resealed (6)

15   The need for energy saving can be quite illuminating for this in Approved Document L (8)

Down

2   Cut one of these in a joist and it must be in the right place (4)

3   The lowest area to be considered for ventilation in a dwelling (8)

4/13 We use this to provide ventilation (7,3)

5   To provide fire protection we may need to install a fire alarm ......? (6)

7   This type of material may be used to reinstate a fire barrier (11)

8   A hole through this must be correctly located (5)

9   One area where we can park up without worrying about ventilation (6)

11   As well as fire this is the one to make a noise about not spreading (5)

13   See 4

Part 3

There are two approved documents which we still need to consider: Approved Document 'M' Access and Use and Approved Document 'P' Electrical Installations.

Access to and use of buildings

Approved Document M requires reasonable provision for people to gain access to and have use of a building and the facilities inside.

It is intended to help those people who have limited reach or mobility to use the building and the facilities. It refers to the requirement for access both into and inside the building such as ramps, lifts and doorways together with the provision and location of equipment such as sinks and toilets.

The electrical installation is included in this requirement by the location of equipment such as socket outlets and switches.

This requirement applies if:

- a non-domestic building or a dwelling is newly erected

- an existing non-domestic building is extended, or undergoes a material alteration

- an existing building or part of an existing building undergoes a material change of use to a hotel or boarding house, institution, public building or shop.

Approved Document M does not apply to an extension or a material alteration of a dwelling. However, an extension or a material alteration of a dwelling must not make the building any less satisfactory in respect to the requirements of Part M than it was before. So a rewire of a dwelling is not necessarily subject to compliance with Part M. Should it be known that the dwelling is to be used or occupied by persons whose reach is limited then adoption of the Part M requirements for socket outlets, switches and the like should be considered.

The location of socket outlets and switches etc, as recommended in Section 8 of the Approved Document, are shown in Figure 2.17.

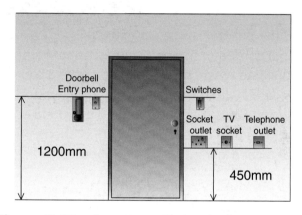

Figure 2.17 *Location of electrical accessories. All switches, sockets and the like are to be mounted between 450mm and 1200mm from floor level*

Electrical safety

Approved Document P relates to electrical installations in dwellings. It requires that:

> Reasonable provision shall be made in the design and installation of electrical installations in order to protect persons operating, maintaining or altering the installations from fire or injury.

The requirements of this Approved Document apply only to electrical installations that are intended to operate at low or extra-low voltage and are:

- within or attached to dwellings such as houses and flats
- in the common parts of a building serving one or more dwellings, common access areas in blocks of flats such as corridors and staircases, shared amenities including laundries and gymnasiums, but excluding power supplies to lifts
- in a building that receives its electricity from a source located either within or shared with a dwelling. This includes dwellings and business premises that have a common supply such as shops and public houses with a flat above
- in a garden or in or on land associated with a building where the electricity is from a source located within or shared with a dwelling. This includes fixed lighting in gardens, pond pumps and outbuildings such as sheds, detached garages and greenhouses.

It is a statutory requirement that the electrical installation work not **excluded** in Approved Document P is notified to the local authority building control. This may be done by either:

1 Notification being given to the local authority building control office prior to the work commencing. The building control authority (or their representative) may then inspect during construction and upon completion of the work. If all is OK then the building control authority will issue a Building Regulations Compliance Certificate to the client.
Or
2 Registration with one of the government accredited Scheme Operators. Once accepted,

the electrical contractor notifies the Scheme Operator when the work has been carried out. The Scheme Operator then notifies the Local Authority and issues the Building Regulations Compliance Certificate to the client.

Remember
The Building Regulations Compliance Certificate and the electrical certification are two separate requirements. Electrical certification must be produced by the electrical installer for all work, whether it is notifiable or not. The building regulations compliance certificate declares that the work complies with all the relevant building regulation requirements, not just Approved Document P.

Work that need not be notified to building control

As mentioned earlier Approved Document P identifies work which is not notifiable and thus implies that all other work is to be notified. The excluded work is given in Table 1 of the Approved Document and the types of work include work which consists of the:

- replacement of the cable for a single circuit only, which has been damaged by rodents, impact or fire, providing the replacement cable has the same current-carrying capacity and follows the same route
- replacement of any fixed electrical equipment (for example, socket-outlets, control switches and ceiling roses), providing the work does not include the provision of any new fixed wiring
- installation or upgrading of main or supplementary protective bonding

- provision of mechanical protection to existing fixed installations, providing the circuit's protective measures and current-carrying capacity of conductors are not affected by increased thermal insulation

- re-fixing or replacing the enclosures of existing installation components, providing the protective measures for the circuit are unaffected

- electrical installation work that is not in a kitchen, a special location, does not involve a special installation and consists of:

  - adding lighting points to an existing circuit including light fittings and switches

  - adding socket-outlets and fused spurs to an existing ring or radial circuit.

  This is relevant if the existing protective device is suitable and provides protection for the modified circuit and other relevant safety provisions are satisfactory.

In addition work not in a special location which is carried out on:

- prefabricated equipment sets and associated flexible leads with integral plug and socket connections

- telephone or extra-low voltage wiring and equipment for the purposes of communications, signalling, information technology, controls etc.

is not notifiable.

Table 2 in Approved Document P identifies the special installations and locations and these include:

Special locations

- swimming pools or paddling pools

- locations containing a bath tub or shower basin

- hot air saunas.

Special installations

- solar photovoltaic (PV) power supply systems

- small-scale generators such as microCHP (micro combined heat and power) units

- electric floor or ceiling heating systems

- garden lighting or power installations

- extra-low voltage lighting installations, other than pre-assembled, CE-marked lighting sets.

Figure 2.18 *Typical bathroom installation*

> **Note**
> BS 7671 and IET Guidance Note 7 give more guidance on achieving safe installations where people are at greater risk.

Much of the information contained in Approved Document P relates to methods of constructing an installation which is safe for use, maintenance and inspection and testing. The most straightforward method of ensuring compliance with the requirements of Approved Document P is to construct the installation to comply with BS 7671. For installations rated up to and including 100 A the installation may alternatively be constructed in accordance with the on-site guide.

As we have seen there is quite a lot of information in the Approved Documents of the Building Regulations. Each one is intended to provide guidance on how to meet the statutory requirements of the Building Act and the Building Regulations. Many of the requirements are related to energy efficiency and care of the environment.

The Code for Sustainable Homes

There is one further requirement that we must be aware of and that is the Code for Sustainable Homes. This was published in 2007 and relates to new homes which are to be rated in relation to their sustainability and to be certified for all new properties as part of the Home Information Packs (HIPs).

So what is the Code for Sustainable Homes?

The Code is the national standard for the sustainable design and construction of new homes. It aims to reduce carbon emissions and create homes that are more sustainable.

The sustainability of a new home is measured against categories of sustainable design, and the rating is for the complete home as a total package. The Code uses a 1 to 6 star rating system to identify the overall sustainability performance of a new home. The Code sets minimum standards for energy and water use at each level.

The criteria by which the sustainability of new homes are measured are:

- energy and CO_2 emissions
- water H_2O and surface water run-off
- materials
- waste

- pollution
- health and well-being
- management
- ecology.

The code requires levels of performance to be achieved for the dwelling and the basis of the sustainability rating is shown in Figure 2.19 below.

Achieving a sustainability rating

| Code Level | Minimum Standards | | | | Other Points[4] Required |
| | Energy | | Water | | |
	Standard (percentage better than Part L[1] 2006)	Points Awarded	Standard (litres per person per day)	Points Awarded	
1(★)	10	1.2	120	1.5	33.3
2(★★)	18	3.5	120	1.5	43.0
3(★★★)	25	5.8	105	4.5	46.7
4(★★★★)	44	9.4	105	4.5	54.1
5(★★★★★)	100[2]	16.4	80	7.5	60.1
6(★★★★★★)	A zero carbon home[3]	17.6	80	7.5	64.9

Notes

1. Building Regulations: Approved Document L (2006) – 'Conservation of Fuel and Power.'
2. Zero emissions in relation to Building Regulations issues (i.e. zero emissions from heating, hot water, ventilation and lighting).
3. A completely zero carbon home (i.e. zero net emissions of carbon dioxide (CO_2) from **all** energy use in the home).
4. All points in this document are rounded to one decimal place.

© Department for Communities and Local Government

Figure 2.19 *The minimum standard and number of points required in order to achieve each level of the code*

Figure 2.20 on the following page summarizes all of the minimum standards which exist under the code.

The term potable water refers to water which is fit for consumption by humans and other animals and is often called drinking water reflecting its intended use. Some water may be naturally potable such as pure natural springs, whilst other sources may need to be treated in order to be safe. The safety of water is assessed using tests which look for potentially harmful contaminants.

Minimum standards		
Code Level	**Category**	**Minimum Standard**
	Energy/CO$_2$	
1(★)	Percentage improvement over	10%
2(★★)	Target Emission Rate (TER)	18%
3(★★★)	as determined by the	25%
4(★★★★)	2006 Building Regulation	44%
5(★★★★★)	Standards	100%
6(★★★★★★)		A 'zero carbon home' (heating, lighting, hot water and **all** other energy uses in the home)
	Water	
1(★)	Internal potable water	120 l/p/d
2(★★)	consumption measured in	120 l/p/d
3(★★★)	litres per person per day (l/p/d)	105 l/p/d
4(★★★★)		105 l/p/d
5(★★★★★)		80 l/p/d
6(★★★★★★)		80 l/p/d
	Materials	
1(★)	Environmental impact of materials†	At least three of the following 5 key element of construction are specified to achieve a BRE Green Guide 2006 rating of at least D – Roof structure and finishes – External walls – Upper floor – Internal walls – Windows and doors
	Surface Water Run-off	
1(★)	Surface water management	Ensure that peak run-off rates and annual volumes of run-off will be no greater than the previous conditions for the development site

Minimum standards *(continued)*		
Code Level	**Category**	**Minimum Standard**
	Waste	
1(★)	Site waste management	Ensure there is a site waste management plan in operation which requires the monitoring of waste on site and the setting of targets to promote resource efficiency
	Household waste storage	Where there is adequate space for the containment of waste storage for each dwelling. This should allow for the greater (by volume) of the following EITHER accommodation of all external containers provided under the relevant Local Authority refuse collection/recycling scheme. Containers should not be stacked to facilitate ease of use. They should also be accessible to disabled people, particularly wheelchair users and those with a mobility impairment OR at least 0.8m^3 per dwelling for waste management as required by BS 5906 (Code of Practice for Storage and On-site Treatment of Solid Waste from Buildings)

Figure 2.20 *Summary of minimum standards*

© Department for Communities and Local Government

Grey water is the waste water produced from domestic activities such as washing, laundry, bathing and the like which can be recycled. This does not include Blackwater (or sewage) which contains human waste.

Improvements in water consumption may be achieved by rainwater harvesting and grey water usage for watering plants, flushing toilets, etc.

> **Note**
> Approved Document G1 of the Building Regulations refers to 'wholesome water' which is the same as potable water. Approved Document G further states that wholesome water must be provided where any water is drawn off to any wash basin, fixed bath or shower, bidet and any sink where food is prepared.

The Government announced the suspension of Home Information Packs (HIPs) with immediate effect from 21 May 2010. The requirement for sellers to give either a Code for Sustainable Homes Certificate or a nil-rated certificate (a sustainability certificate) to the buyers of newly constructed homes has also been suspended.

However, the Code for Sustainable Homes is still operational and remains the Government's national sustainability standard for new homes.

> **Note**
> It is important that you check the status of the HIPs and the Code for Sustainable Homes regularly as these may be reintroduced at any time.

Having considered the requirements of the Building Regulations and the Code for Sustainable Homes complete the Task and the Self Assessment Questions before moving on to the next chapter.

Task

Look around your home or workplace and identify one area where the requirements of each of the following Approved Documents would apply.

- A: Structure

- B: Fire safety

- E: Resistance to the passage of sound

- F: Ventilation

- L1: Conservation of fuel and power in dwellings

- M: Access to and use of buildings

- P: Electrical safety

SELF ASSESSMENT

Circle the correct answers.

1   Approved Document M of the Building Regulations applies to:
a. All dwellings
b. Existing dwellings only
c. All new dwellings
d. All alterations to dwellings

2   The minimum height for an electrical accessory to comply with Approved Document M is:
a. 300mm
b. 350mm
c. 400mm
d. 450mm

3   The type of buildings to which Approved Document P of the Building Regulations applies are:
a. Public
b. Dwelling
c. Commercial
d. Industrial

4   Which of the following electrical works would require notification to local authority building control?
a. The installation of a replacement electric shower unit
b. The installation of two new sockets on the bedroom ring circuit
c. The installation of a replacement damaged cable in a lighting circuit
d. The installation of a new circuit supplying a cooker

5   The term potable water means:
a. drinking water
b. waste water
c. sewage water
d. rain water

Classification
of materials

RECAP

Before you start work on this chapter, complete the exercise below to ensure that you remember what you learned earlier.

The Building Regulations Approved Document which identifies the requirements for Fire Safety relating to buildings other than dwellinghouses is?

_____

One area of a commercial building where people normally go but which would not require ventilation is?

_____

Approved Document L identifies that for external lighting the lamp capacity should not exceed 150W and be?

_____

Approved Document M states that the minimum height for a socket outlet in a location where it may be used by someone with limited reach is?

_____

Is the addition of two socket outlets to a ring final circuit within the bedroom of a dwelling notifiable under Approved Document P?

_____

LEARNING OBJECTIVES

On completion of this chapter you should be able to:

● State materials and products classified as:

  – hazardous to the environment

  – recyclable.

● Describe organizational procedures for processing materials classed as:

  – hazardous to the environment

  – recyclable.

Part 1   Classification of materials

The materials and products that we use in the construction of a building and the electrical installation all have some impact on the environment. We are most concerned with establishing whether they are:

● hazardous to the environment
● recyclable.

Hazardous materials

Hazardous materials are those which contain hazardous substances in quantities which are liable to cause:

● death or injury to people and wildlife
● pollution of waters
● pollution of air

● unacceptable impact on the environment if not properly dealt with.

These materials need to be handled and stored with care and any waste disposed of safely.

Materials that are commonplace in the electrical installation and construction industries which may be classed as hazardous materials include:

● Asbestos: This is found in a number of older building materials and equipment and includes artex ceiling finishes, spray covering insulation/fire protection on structural steelwork and ceiling and wall tiles. It was also included in electrical components and equipment, for example asbestos pads were common in old, ceramic, rewirable (BS 3036) fuse carriers and holders.

Figure 2.21 *Asbestos in old fuseboards*

● Batteries: Includes small batteries used in test instruments up to the large lead acid cells used for emergency lighting installations. Many of these lead acid cells have been replaced with gel batteries.

Figure 2.22 *All batteries are hazardous but can be recycled*

● Discharge lamps: Fluorescent lamps have been widely used in commercial and industrial installations for many years. Low energy compact fluorescent lamps are being used for many more lighting applications and contain low pressure mercury vapour which must be disposed of carefully. Sodium vapour discharge lights, commonly used in street lighting, stadia and areas where colour rendition is not essential are also subject to the same requirements.

● Adhesives: Many adhesives are hazardous and will be recorded under the COSHH requirements. Electricians frequently use adhesives when installing PVC.

● CFCs and HCFCs: These are chlorofluorocarbons (CFCs) and hydrochlorofluorocarbons (HCFCs). CFCs are organic compounds that contain carbon, chlorine and fluorine. Hydrochlorofluorocarbons (HCFCs) contain carbon, chlorine, fluorine and hydrogen. CFCs were widely used as refrigerants, aerosol propellants and solvents. Their production is being phased out because of their impact on ozone depletion.

● Oils and oil filters.

Other hazardous materials include:

● paints and coatings
● plastics
● toxic metals: common examples are lead pipe work and flashing and mercury in micro-switches
● electronic equipment
● sewage
● TVs and radios
● computers, computer screens and cathode ray tubes
● rubber and tyres
● glass
● cement and plaster
● thermal insulation.

This list is not exhaustive but gives some indication of the materials that are classed as hazardous. Indeed the Hazardous Waste (England and Wales) Regulations 2005 identify some 91 components of waste that are considered hazardous and 14 properties that render waste hazardous.

Task

The Environment Agency produces the following guidance on hazardous waste: *Environment Agency HWR01: What is a Hazardous Waste?*

Using either library facilities or by going online familiarize yourself with the guidance provided in HWR01.

_____

_____

Recyclable materials

The classification of a material as hazardous does not mean it cannot be recycled. Recycling relates to the nature of the waste material and whether the material or its content can be recovered and reused. The recycling of plastic drinks bottles, for example, supports the delivery of numerous fleece jackets onto the market following the recycling process.

Modern methods have greatly improved both the recycling and the materials which may be recycled. Manufacturers have moved towards sustainable and recyclable materials and the reduction in packaging material for goods is on the increase.

In truth many of the materials used are now recyclable so let's consider a few of those we may come across.

Paper and card: These common packaging materials, by their nature, do not generally constitute an environmental hazard. They may create other hazards such as fire, rodent infestation and obstruction if incorrectly located or stored. Paper and card is widely recycled and these materials should be segregated on site and sent for recycling.

Metals (steel conduit, trunking, ducting etc.): Most metals are recyclable and all waste and redundant metal items should be recycled. Some metals do present an environmental hazard, such as lead pipes, and these should be treated as hazardous waste but still recycled using authorized waste disposal contractors.

Discharge lamps (including compact fluorescent lamps): These are hazardous but are recyclable. The lamps should not be broken on site and require storage in a designated container for collection by an authorized handler. The materials within the lamps are recyclable but only by authorized contractors using specialist processes.

Plastics: There are many types of plastic and much of it is recyclable. However, there are certain types being manufactured, some of which are used for packaging, which are not suitable for recycling. Separation of the recyclable plastic should be carried out so that it may be recycled and the non recyclable sent to landfill waste. Many manufacturers are using biodegradable materials nowadays which, whilst not suitable for recycling, break down harmlessly over time. There are many products which are constructed using plastic such as conduit and trunking, equipment casings and cabinets and appliances.

Electrical equipment will need to be dealt with in accordance with the WEEE Regulations. Details of the conduit and trunking and other plastics would need to be sourced from the manufacturer before disposal.

Handling materials

So how do we go about dealing with these materials on site? There are two main elements involved in this process:

● materials used on site during the construction process

● waste materials both during construction and on completion.

Materials during construction

Materials which are hazardous to health must be recorded in the COSHH register, together with the recommendations for their handling and disposal. For example, the adhesives used with plastic containment systems (conduit and trunking) will normally carry warnings about:

● contact with skin, eyes and other materials

● the need to be used in well ventilated areas due to the fumes produced

● avoiding contact with foodstuffs, both directly and by hand contact.

Figure 2.23 *Conduit adhesive is registered under COSHH as a hazardous substance*

Operatives are required to cooperate with their employer to meet the safety requirements and this means following both manufacturers' and industry guidance together with their company procedures for dealing with these materials.

Equipment which is being installed such as cables, containment systems, etc. will also have requirements in regards to their handling, although these may not all relate to specific hazards. For example, the temperature range for the installation of PVC cables should be complied with to prevent damage to the insulation.

However, the handling and termination of fibre optic cables does carry some hazards in respect to:

● glass offcuts getting into the skin and eyes

● small particles which could be inhaled or swallowed

● not looking into the ends of the fibres when testing or whilst they are in operation.

The splicing and termination of fibre optic cables uses adhesives and cleaners as part of the process and it is vital that the instructions for use are followed. For example, even a basic isopropyl alcohol cleaner is flammable.

So we can see that there are a number of issues to be considered during the construction for both the materials and compounds we use.

Waste materials

First we need to consider: what is waste?

One description of waste would be anything not wanted or left over. The Environmental Protection Act provides a description of 'waste' which includes any substance which constitutes:

- a scrap material
- an effluent
- other unwanted surplus substance arising from the application of any process

and any substance or article which requires to be disposed of as being:

- broken
- worn out
- contaminated
- otherwise spoiled.

It does not include any substance which is an explosive.

Having established what waste materials are, we need to consider which waste materials are also hazardous? A brief description of the properties of hazardous waste was given earlier in this chapter.

The term hazardous waste is often used to include:

- toxic waste: having properties liable to affect living things and include heavy metals (e.g. mercury) and cyanide

- clinical waste: this may include both a toxic material and infectious material, human and animal tissue, drugs, syringes, etc. This waste would come under the Controlled Waste Regulations.

It is important to remember that companies are responsible for the waste their work produces right up to the point of disposal.

Remember

The waste you are likely to deal with will fall into one of the following categories:

- waste that is not considered to be hazardous, for example: paper, cardboard, etc.
- waste considered to be hazardous under the hazardous waste regulations, for example: lead acid batteries or fluorescent tubes
- waste that needs to be assessed to find out whether it is hazardous or not, for example: cutting compounds and lubricants.

So how is the waste dealt with? In general terms, irrespective of the type of waste, the process on site will be much the same.

A point on site should be designated where waste material is to be stored. In the short term, during the working day, smaller containers may be used to collect waste material to be transported to the storage area.

Within the storage area there should be suitable designated storage facilities for each type of waste which is produced and the waste material should be placed in the appropriate area.

The waste should then be collected by an authorized waste disposal contractor, licensed for the specific materials being disposed of or recycled.

Contractors who are licensed to dispose of hazardous waste normally provide suitable receptacles for use on site where the quantities of waste generated warrant it. So, for example, during the refurbishment of commercial premises there is likely to be a considerable number of redundant luminaires (light fittings) many of which will contain fluorescent tubes. In this case a separate storage will be required for the tubes and another for the fittings.

© Balcan Engineering Ltd

Figure 2.24 *Disposal of discharge lighting*

When working on smaller contracts it may be necessary for the individual contractor to dispose of the waste material they produce. These should still be segregated and taken to a suitable recycling and disposal centre. These centres are often run by, or on behalf of, the local authority and they have specific areas for hazardous and recyclable waste and an area for safe 'landfill' general waste.

These facilities make a charge for recycling commercial waste (generated by companies) but waste from the general public within the local authority area can usually be disposed of free of charge.

SELF ASSESSMENT

Circle the correct answers.

1   One of the hazardous materials which may be found when dealing with old porcelain rewireable fuses is:
   a.   lead
   b.   asbestos
   c.   copper
   d.   rubber

2   Non-hazardous waste materials include:
   a.   paper
   b.   lead
   c.   PVC
   d.   batteries

3   One of the essential requirements when working with adhesives is to ensure the area is:
   a.   dry and warm
   b.   not ventilated
   c.   well ventilated
   d.   secure

4   Heavy metals include:
   a.   lead
   b.   copper
   c.   tin
   d.   mercury

5   The electrical contractor is responsible for the disposal of recyclable waste:
   a.   until it is collected from the site
   b.   up to the point of disposal
   c.   up to placing in the site disposal area
   d.   until the end of the contract

2.4

Minimizing environmental impact

RECAP

Before you start work on this chapter, complete the exercise below to ensure that you remember what you learned earlier.

Hazardous materials are those which contain hazardous substances in quantities which are liable to result in which **four** outcomes?

_____

_____

_____

_____

The hazardous material likely to be found in old fuseboards with porcelain fuses is?

_____

Batteries do represent a _____ to the environment and should always be

_____.

Adhesives used for sticking PVC conduit and trunking are _____ and should be registered In the _____ register.

Paper and card are classed as _____ waste products.

Waste materials should be sorted into their relevant _____ and disposed of by _____ handlers.

LEARNING OBJECTIVES

On completion of this chapter you should be able to:

● State methods that can help reduce material wastage.

● Explain the importance of reporting environmental hazards.

● Specify environmentally friendly products, materials and procedures.

Part 1

Having considered the requirements of legislation and the effect of various materials on the environment, what steps can we take to minimize our impact?

The first thing we can do is to employ installation methods which help to reduce material waste. There are many ways in which we can do this so let's consider some of these starting with planning the installation.

Figure 2.25 *Planning the installation*

Planning the installation

By careful planning it is possible to reduce the waste produced from the installation process.

The things to consider are as follows:

The materials to be used: This involves planning the route for the installation before we begin. We would start by deciding on the most suitable route for the electrical containment systems and cables, taking into account the building layout, structural features and other services and equipment. This route should allow for ease of installation, the material used to a minimum.

During the course of the construction the skill of the installer is vital in keeping material waste to a minimum. Accurate installation of containment systems, such as conduit and trunking is essential to reduce waste. Bending conduit accurately to follow the building contours and changes of direction will reduce the cutting and offcuts produced. Cables placed accurately in position with sufficient material left to allow termination will minimize the waste material produced by cable offcuts.

Material management and storage: We need to ensure that materials for the work are delivered and stored to minimize damage.

Damaged or contaminated material is destined for the waste/recycle process and so careful management will minimize such waste.

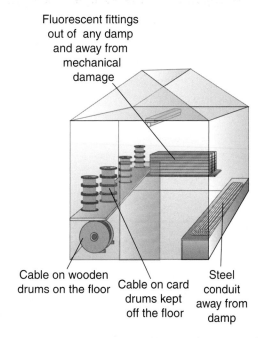

Figure 2.26 *Suitable storage will minimize waste*

Appropriate installation: By ensuring the materials are installed at an appropriate time in the construction process we can minimize damage due to environmental conditions and other work activities. Correct installation together with adequate protection will help to reduce such damage. For example, applying capping over flat twin and CPC cables on walls that are to be plastered will protect them from damage by the plasterer's trowel. There are also products on the market to help minimize the damage to conductors and accessories during the construction process, which may be beneficial.

OPERATION DESCRIPTION	DAY NO								
	1	2	3	4	5	6	7	8	9
1ST FIX INSTALLATION	▬	▬	▬						
2ND FIX INSTALLATION				▬	▬	▬	▬		
INSPECTION AND TEST								▬	
HANDOVER									H/O ◇
DELIVER 1ST FIX	◇								
DELIVER 2ND FIX			◇						

Figure 2.27 *Simple bar chart plan for an installation*

Remember

Damage to cables, accessories and equipment costs money and produces additional waste. Failing to plan the installation and determine material requirements may also result in additional waste.

Task

Using trade publications and tool/equipment manufacturers' catalogues identify **three** devices which will improve installation techniques or provide protection.

1 _____

2 _____

3 _____

Reporting environmental hazards

Should the worst happen and there is an environmental hazard caused by or arising from the work activities it is essential that this is reported promptly. In the first instance, and dependent upon the nature of the hazard, this would be reported to your supervisor or the site manager. Where this does not put you at any further risk efforts should be made to control the hazard.

For example, the discovery of a leak from a diesel storage tank could be minimized by spill control barriers and a receptacle for the leaking oil in the short term. The ground onto which the diesel fuel has spilt should be removed and disposed of in accordance with the appropriate legislation.

More serious hazards may require the involvement of specialist treatment and disposal companies. Serious environmental hazards should be notified to the local authority and may require major intervention and treatment. For example, if a diesel storage tank were to rupture and the spillage pollute a water course then there are serious environmental implications and urgent action is required.

© Reproduced with Permission. JSP Ltd.

Figure 2.28 *Typical small oil spill kit*

Environmentally friendly products, materials and procedures

The requirement to use environmentally friendly materials, products and procedures is resulting in many advances in the materials we use. Whilst the eco-friendly new build property is becoming ever more carbon neutral, existing properties are not so easily converted. In many cases the addition of better insulation, energy saving appliances and the like can all contribute.

The development of improved materials is an ongoing activity and whether selecting or specifying materials we should endeavour to use environmentally friendly products. This may be quite a simple task such as using a water soluble cutting paste rather than an oil-based one, which will provide the same function with less environmental impact. In the same vein there are water reducible resins which may be used in the production process.

The sustainability of concrete products can be improved by the inclusion of recycled aggregates such as crushed concrete, recovered from demolition sites, in concrete mixes. Products from power generation such as fly ash may also be used in construction products such as building blocks, both recycling and improving the carbon footprint. Steel reinforcing in concrete can provide lighter and smaller foundations with the added bonus of recyclable steel at the end of the building's life.

Selecting low smoke and fume and cable and containment systems produced to be environmentally friendly will also improve the sustainability of the building. These products, as the name suggests, produce much lower smoke and fumes in the event of a fire. This in turn reduces the impact on the environment due to the reduction in airborne chemicals.

© Image provided by www. centralcable. co.uk

Figure 2.29 *Low smoke and fume cable*

The use of low energy lighting has already been discussed and this, coupled with making the most use of the natural daylight as the main source of lighting for an interior space has many benefits, including reducing both running costs and carbon footprint.

Changes to traditional heating methods also offer benefits. Underfloor heating for example requires a much lower temperature for the heating element (electric or water) and thereby reduces the environmental impact whilst retaining a comfortable environment.

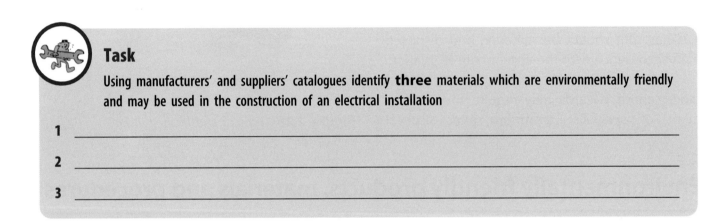

Task

Using manufacturers' and suppliers' catalogues identify **three** materials which are environmentally friendly and may be used in the construction of an electrical installation

1 _____

2 _____

3 _____

SELF ASSESSMENT

Circle the correct answers.

1 Waste on site can be reduced by:

   a. working quickly

   b. careful planning

   c. using short cuts

   d. using sub-contractors

2 To prevent unnecessary damage and waste, materials for the work should be:

   a. stored outside the site

   b. delivered at the beginning of the job

   c. delivered as they are required

   d. stored on site until needed

3 Damage to PVC cables and hence wastage can be reduced by making sure cables are only installed in:

   a. the appropriate ambient temperature

   b. metal trunking and conduit

   c. plastic trunking and conduit

   d. deep wall chases

4 In order to minimize the possible impact on the environment any oil leakage should be reported:

   a. in writing

   b. at the end of the day

   c. at the site meeting

   d. immediately

5 Minimizing the impact on the environment in the event of a fire may be achieved by using cables that have insulation which is:

   a. polyvinyl chloride

   b. low smoke and fume

   c. vulcanized india rubber

   d. cross link polyethylene

2.5

Environmental technology systems

RECAP

Before you start work on this chapter, complete the exercise below to ensure that you remember what you learned earlier.

Careful planning helps to _____ the waste produced during the

_____ process.

The materials for the job should be _____ and stored to _____ damage.

If an environmental hazard occurs due to the work activity this should be

_____.

The sustainability of a building can be _____ by the installation of

_____ friendly _____ and materials.

LSF cables have less _____ on the environment due to the _____ in

airborne _____.

LEARNING OBJECTIVES

On completion of this chapter you should be able to:

● Describe the basic operating principles of environmental technology systems.

● State the applications and limitations of environmental technology systems.

● Identify local authority building control requirements for the installation of environmental technology systems.

Part 1 Environmentally friendly installations

The selection of environmentally friendly materials and products is a vital part of the electrical installation planning and construction process. Companies are required to reduce their own carbon footprint and that of the installations they install, both during construction and throughout the lifetime of the installation.

Part of this process includes the use of 'renewable systems', so called because they are not reliant upon a finite source such as coal, oil or gas.

We do need to understand these systems and their basic operating principles together with their applications and limitations. We will start by considering solar photovoltaic systems.

Solar photovoltaic systems

Photovoltaic cells, generally referred to as PV cells, convert light into electrical energy. However, the output from a single PV cell is quite small so they are connected in arrays to create a 'solar panel' with a higher output. There are a number of panel technologies in use including thin-film, monocrystalline and multicrystalline.

So how does it work?

Figure 2.30 *The photocell*

The basic operation

A PV cell uses similar semi-conductor materials to those used in the microelectronics industry, such as silicon. A thin wafer of the material is treated to produce an electric field with one side positive and the other negative. Light striking this cell causes electrons to be dislodged from the atoms of the semiconductor material and if conductors are attached to the positive and negative sides of the cell then an electric current flows as a result. The amount of current produced is directly proportional to the amount of light striking the cell.

Connecting these cells in series will increase the voltage output and connecting them in parallel will increase the current flow.

We can see from this that the output from these cells will be direct current and so this may be used to supply items of equipment operating at dc, such as mobile phones and laptops, garden lighting and caravans. Where PV arrays are used to provide energy for ac electrical installations then their output is connected to a static inverter to convert it from dc to ac.

Figure 2.31 *Typical small PV unit for garden lighting*

Small groups of cells can be used to power smaller items as described above, larger arrays can produce kilowatts (kW) of output suitable for use in domestic dwellings. Much larger arrays may be used to supply commercial and industrial units or solar power plants which supply energy directly to the electricity distribution system.

Typically, a domestic system rated around one kilowatt (1kW) will cover an area of approximately ten square metres. Such a unit may produce around 750kWh a year. The bulk of this output will be in the summer time.

In the UK the majority of such systems are south facing and roof mounted. Ground located systems are more often used in locations where ground space is not critical and therefore not as common in the UK.

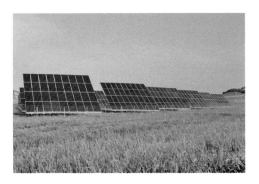

Figure 2.32 *Typical ground located PV system*

The energy produced by the PV system is used to power the installation and when there is more energy produced than is used by the installation, the surplus energy may be sold to the distribution network operator.

Some of the advantages and limitations are given in Table 2.1.

Table 2.1 *PV systems*

Advantages	Limitations
Renewable energy	Initial outlay is expensive
Costs nothing to produce	Long cost recovery period
May reduce energy bills (reduced consumption and sell back)	Only operable in daylight
Useful in remote areas where other supplies may be difficult to provide	Output will vary on the strength of light available
Low maintenance (no moving parts)	

PV arrays are often combined with micro wind generation to provide a more reliable energy supply. The PV array will provide energy in the summer, as this is generally when winds are low and sunlight is plentiful and strong. Wind generation provides energy in the winter as this is generally when sunlight is of short duration and weak whilst the wind is generally stronger.

This type of arrangement is often seen on traffic signs, weather stations and the like. These are particularly useful in remote locations and for use in temporary locations such as road works.

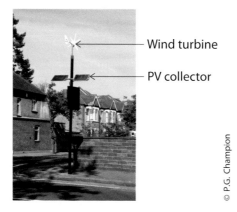

Figure 2.33 *Combined PV and wind turbine traffic information sign*

Wind generation (micro and macro)

Wind generation is used for small-scale (micro) and large-scale (macro) generation. Micro generation includes the small generation units mentioned above and the wind generators used for dwellings and industrial and commercial buildings. Macro generation relates to the larger wind turbines used for wind farms and supplying electrical energy into the transmission and distribution system. Midi or small turbines are generally those in the region of 2–8kW.

How do they work?

The basic operation of the wind turbine is the same whether it is micro or macro generation.

The wind is a natural resource. A rotor blade connected to a shaft and placed in the direction of the wind converts the wind energy into rotational mechanical energy. The shaft is connected through a gearbox to a generator which converts the rotating mechanical energy into electrical energy. Put simply, it does this by rotating permanent magnets within a coil of wire, which produces an alternating current. The whole assembly, including some control gear, is mounted in an enclosure called a nacelle. The conductors carrying the current produced pass down the hollow shaft which supports the nacelle. The direction the blades face to turn them into the wind may be controlled by a small vane for the smallest micro generators, moving up to electronic control gear and directional motors, often referred to as 'yaw' motors, for the larger macro generators.

Figure 2.34 *Micro wind generator*

Figure 2.35 *Macro wind generation*

The majority of wind turbines in use today are horizontal shaft turbines, which look like aeroplane propellers. There are also vertical shaft turbines, which generally have curved vertical blades between rings top and bottom. Irrespective of the type the principle is the same.

Figure 2.36 *Vertical axis wind turbines*

Table 2.2 *Wind generation systems*

Advantages	Limitations
Micro/Midi	
Renewable	Installation cost can be high
Reasonable maintenance cost	Not suitable for low wind areas
Good in higher wind areas	Moving parts so maintenance required
Sell back with midi generators	Above 2kW before sell back (usually)
Macro	
Renewable	High wind areas
Ideal in remote locations	Impact on landscape
Comparatively low maintenance	

© Photos courtesy of Windspire Energy www.WindspireEnergy.com

The essence of their operation is to have the turbine propeller at a sufficient height and location to catch and use the wind energy. On wind farms there are often large dominant support shafts and large blades to achieve the appropriate height and maximum effectiveness. They also have pitch control motors which alter the angle of the propeller blade into the wind.

Some of the advantages and limitations are given in Table 2.2.

Remember

A small increase in the wind speed will lead to a much larger increase in the amount of energy generated. In fact, the change is cubed and so half the wind speed will generate an output only an eighth as much whilst twice the wind speed will lead to eight times the output. As wind speeds increase with height it is normally considered best to mount micro and mini turbines on a mast.

Try this

1   Photovoltaic cells are connected in _____ because the output from a single cell is very _____. The output from the cells is a _____ voltage and must be passed through a _____ to provide an output suitable for an ac electrical installation.

2   List one advantage and one disadvantage for a PV system.

   a.   Advantage

   _____

   b.   Disadvantage

   _____

3 Directing the _____ of a wind turbine into the wind may be by use of a small _____ for micro systems and a _____ for macro systems.

4 The two types of blades used for wind turbines are _____ and
_____

5 List one advantage and one disadvantage for a wind turbine system.

a. Advantage

_____

b. Disadvantage

_____

Part 2

Hydro-electric power

The most widely used natural resource for generating electricity is hydro-electric power. In some countries where lakes and reservoirs are plentiful the electricity supplies are almost entirely generated by hydro-electric plants. Although electricity is cheap to produce in this way it often has to be transmitted over long distances to the nearest populated and industrial areas. To produce electricity using hydro-electric power a head of water has to be available. Figure 2.37 shows how reservoirs high in the mountains are used to create pressure through an artificial tunnel.

Hydro-electric power stations are used to produce considerable output and with a fast response time when the demand arises. For example, the hydro-electric power station at Ffestiniog in Wales can produce 360 megawatts and reach this level of output within 1 minute should it need to do so.

We also have to look at the smaller hydro-electric power sources.

Micro hydro generation

The term micro hydro is used for hydro-electric generating systems which produce up to about 100kW output. These can be used to supply dwellings or small industrial/commercial installations, particularly those in remote areas.

Pico hydro generation

The term pico hydro generally relates to hydro generating sets producing an output up to about 5kW. These are suitable for small-scale use and the smaller units (less than 1kW) quite often rely

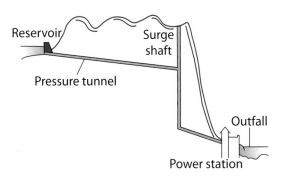

Figure 2.37 *A typical hydro-electric station*

on the natural flow of a river. Water is taken from the river using a pipe, then passed down a slope and through the turbine. The water is then piped back into the river further downstream.

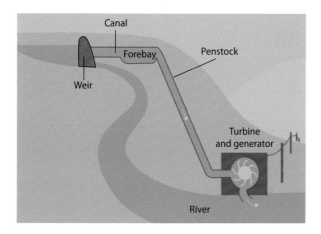

Figure 2.38 *Typical micro hydro scheme*

The simplest form of hydro generation uses a waterwheel to drive a generator; however, as the rotation of the waterwheel is likely to be quite slow some form of gearbox is required. There are a number of generators produced which use propellers to drive the generator and this may, for the smaller units, be placed directly in the flow of water, providing the flow rate is high enough.

Some of the advantages and limitations are given in Table 2.3 below.

Table 2.3 *Hydro-electric systems*

Advantages	Limitations
Cheap to produce (no fuel costs)	Has to be located near suitable water source
Uses a natural renewable resource	Affected by water flow rates
Reliable	Effect on landscape (where dams are required)
	Effect on natural water flow rate

Having considered the three most noticeable forms of renewable energy we need to be aware of some of the less obvious systems, beginning with heat pumps.

Heat pumps

A heat pump, put simply, is a device which moves heat from one location, the source at a lower temperature, to another location, the heat sink, at a higher temperature.

There are two main types:

- ground source heat pumps (GSHP)
- air source heat pumps (ASHP).

Their principle of operation is the same: they extract heat from the ground or the air and transfer this to the building, rather like a fridge in reverse.

How do they do this? Well the GSHP, for example, takes low-grade heat from the ground. At a depth below one metre there is a stable source of heat throughout the year. This low-grade heat is then passed through an evaporator to the refrigerant in the heat pump. The refrigerant is then compressed, by an electrically driven compressor, which greatly increases its temperature. This high-grade heat is then passed to the building to provide heating and hot water. This is the same principle that a refrigerator operates on, except the fridge reduces the temperature in the cabinet and produces heat as a waste product. The heat pump operation for the ASHP is essentially the same as that for the GSHP.

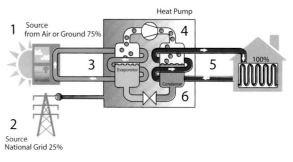

Figure 2.39 *Basic heat pump principle*

The ground source heat collector can be either laid shallow (1m or more) over a large area or deep in a bore hole, which will require specialist drilling equipment.

Whilst the heat pump does require an electrical supply, the efficiency of the pump is around 300–400% so the heat output is considerably higher than the energy used. For every 1kW of energy used to operate the heat pump the unit produces 3–4kW of heat energy for the installation.

The type of heat produced is lower than that of a conventional boiler and is ideally suited to underfloor heating. There are special radiators produced for ordinary wet systems; or traditional radiators will need to be larger to provide the heat output. An immersion heater may be fitted in the water storage tank to provide a boost for hot water when demand dictates, but the whole system is considerably more efficient than other traditional heating systems.

Some of the advantages and limitations are given in Table 2.4 below.

Table 2.4 *Heat pumps*

Advantages	Limitations
Highly efficient	Does require a power supply
Uses a natural renewable resource	Additional changes required for retrofit
Reliable	Does not provide electrical energy
Constant heat supply	
Quiet operation	
Unobtrusive	
Supplies heating and hot water	

Try this

1 Hydro-electric generation is produced using a generator and _____ driven by water. This may be supplied from a _____ or from free _____ water such as rivers.

2 The main component parts for a hydro scheme using a river are the:

a. F_____

b. P_____

c. T_____

d. G_____

3 The two most common types of heat pump are the _____ source and the _____ source. They operate by taking heat from the _____ and transferring it to the _____ using a unit rather like a _____.

4   A heat pump does use electrical energy but the efficiency is high being at least:

   a.   90%

   b.   100%

   c.   200%

   d.   300%

5   List three advantages of a heat pump:

   a.   _____

   b.   _____

   c.   _____

Part 3

Combined heat and power

The next system we are going to consider is combined heat and power (CHP) and microCHP.

In simple terms the combined heat and power (CHP) uses the heat produced during the generation of electricity to provide heating and hot water for consumers. The traditional electrical generation system produces heat and this is generally vented to the atmosphere, the steam clouds above power station cooling towers are an example of this.

The CHP uses the heat produced and, rather than wasting it uses it, to provide heating and hot water.

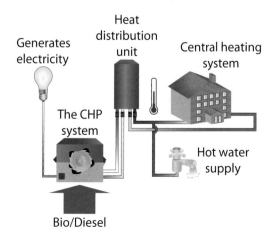

Figure 2.40 *Typical CHP system*

The fuel used for the generation process may be any type of fuel but the use of biodiesel shown in Figure 2.40 would be a suitable option. Even the use of fossil fuels produces a considerably more efficient output due to the utilization of the heat produced.

CHP units may be used to provide supplies for individual buildings such as a school and such a unit may reasonably deliver an electrical supply of 12–15kW and a heating supply of 26kW. This may be powered using the existing natural gas supply replacing the traditional gas boiler. Control panels are used to ensure optimum output levels with minimum waste and a number of units may be connected in tandem to increase the output.

MicroCHP

The microCHP units used for individual dwellings are generally smaller and provide an output in the region of 1kW for use in the home. The units normally produce electricity when heat is required and any electrical energy not used in the property may be sold back to the local network.

These smaller micro units generally will not totally replace the supply from the distribution network operator, although that is dependent on the type and construction. Larger power station versions generate the electricity and use the heat from that process. Many of the microCHP units generate heat and produce electricity as part of that process.

Figure 2.41 *A Baxi Ecogen micro CHP unit*

Some of the advantages and limitations of CHP are given in Table 2.5 below.

Table 2.5 *CHP*

Advantages	Limitations
Does not rely on weather conditions.	
Generates electricity at times when heat is produced	Only provides electrical energy when heat is required
Cost saving	
Increased efficiency	
No planning permission required	
Reduces carbon emissions	

Bio-heating

The next system we are going to consider is the use of bio heating. There are three main types of heating that use bio fuels:

- Bioliquid uses mineral oil such as biodiesel to replace conventional fuel.
- Biogas uses the gas produced from waste material, some from waste treatment works, and others from landfill waste sites where methane and other gases are produced from the decomposition process.
- Biomass, which we are going to consider in a little more detail.

Smaller systems using waste products may be installed for domestic premises.

Biomass heating

The term biomass refers to organic matter which forms the basis of the fuel and includes:

- wood from forest coppicing and wood processing
- agricultural residue from harvesting and processing
- food waste from food manufacture, processing, preparation and leftovers
- crops grown specifically for fuel.

These materials give off carbon dioxide when burnt but this is offset by the carbon dioxide absorbed by the material during their growth due to photosynthesis. The materials are also sustainable and are replanted once they have been harvested and so the cycle begins again.

The most common UK biomass crops are generally short rotation such as coppiced willow and poplar and oil seed rape.

In addition to specifically grown crops, other agricultural byproducts are also referred to as biomass and these include straw, grain husks, forest products, waste wood and animal wastes such as slurry and chicken litter.

All these materials are used in place of the traditional fossil fuels.

Some of the advantages and limitations are given in Table 2.6 below.

Table 2.6 *Biomass*

Advantages	Limitations
Low cost	Installation cost
Virtually carbon neutral	Less easy to control
Utilizes renewable energy	Storage for fuel

Try this

For multiple choice questions circle the correct answer.

1   The term CHP stands for:

   a.   central heating pipes
   b.   carbon heat products
   c.   combined heat and power
   d.   centralized heat and power

2   A CHP unit can run on any fuel and benefits the environment because it uses the waste _____ from the generation of _____ energy to provide _____ and hot _____.

3   A microCHP unit is usually smaller and generally produces an electrical output in the region of:

   a.   0.5kW
   b.   1.0kW
   c.   1.5kW
   d.   2.0kW

4   The main types of bio fuels are bioliquid, biogas and biomass. Give one example of a typical source for each type.

   a.   bioliquid

_____

   b.   biogas

_____

   c.   biomass

_____

5 The fuels specifically grown for biomass can also be supplemented using byproducts from other processes. List three common byproduct materials which may be used.

a. _____

b. _____

c. _____

Part 4

The last of our environmentally friendly systems is solar thermal heating.

Solar thermal heating

We have already considered the use of PV systems for the generation of electrical energy. A solar thermal heating system uses the heat generated from sunlight to heat water. The system uses an indirect cylinder as shown in Figure 2.42 to transfer the heated water within the solar collector to the consumer's hot water.

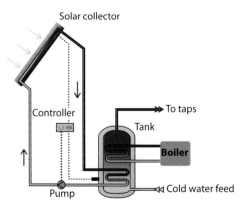

Figure 2.42 *Typical solar water heating system*

One of the drawbacks is that the peak demand for hot water is generally in the winter months when

sunlight is at its weakest. It cannot therefore supply all the hot water needs but could deliver around 50% of the hot water requirements over the year and considerably reduce the cost of hot water.

The solar collectors are generally roof mounted, facing south, and require plumbing in to complete the system. A PV panel may be mounted adjacent to the collector to power the solar heating pump.

Collectors are either flat-profile, which can be more easily incorporated into the roof fabric, or evacuated tube types, which have a round profile and generally have a larger contour than the flat profile type.

Figure 2.43 *Typical evacuated tube solar collector*

Some of the advantages and limitations are given in Table 2.7 below.

Table 2.7 *Solar thermal heating*

Advantages	Limitations
Low running cost	Installation cost
No emissions	Limited availability
Utilizes renewable energy	Long term payback

Water recycling and rainwater

There are a couple of other environmentally sustainable options which we must consider, grey water recycling and rainwater harvesting, both of which relate to water saving.

Grey water recycling

> **Remember**
>
> Grey water is the waste water produced from activities such as laundry, washing up and bathing or showering. It is identified as grey water as it has been used and although it generally contains soaps etc. it can be recycled and used on site. It is distinct from black water which is sewage water containing human waste and this cannot be recycled on site as it requires specialist treatments (see Hazardous Waste and Code for Sustainable Homes covered earlier).

Grey water represents between 50 and 80% of the waste water produced in residences, excluding the toilet waste.

The use of grey water for irrigation and the like has been around for some time. Toilets which utilize the grey water produced in the bathroom allow the grey water to be recycled for flushing toilets.

Figure 2.44 *Grey water toilet flushing system*

Some of the advantages and limitations are given in Table 2.8 below.

Table 2.8 *Grey water recycling*

Advantages	Limitations
Environmentally friendly	Installation cost
Reduces water usage	Storage facilities
Saving on water bills	

Rainwater harvesting

As the name suggests rainwater harvesting is simply collecting, saving and reusing rainwater. In its simplest form this may be a water butt under a downpipe with the water used for watering the garden.

More elaborate systems collect and store the water from roofs and use it not only for irrigation but for the washing machine, toilet flushing and so on.

Large systems can use underground storage vessels and collect rainwater from a large area. This can then be used to supply large irrigation projects such as golf courses, the supply of recyclable water for car washes and so on.

There are two basic types of rainwater harvesting system:

- header tank
- direct pump.

The header tank requires a high-level storage facility, generally at roof level, and the water is supplied under the head of water pressure created by gravity. The system is effective but requires a suitable storage facility with adequate support. Also, toilet cisterns and washing machines may take some time to refill on this system.

Where the water is stored underground or the supply is to be used for washing machines, etc. then a pump will be required to generate sufficient water pressure (equivalent to mains water supply pressure).

Some of the advantages and limitations are given in Table 2.9 below.

Table 2.9 *Rainwater harvesting*

Advantages	Limitations
Environmentally friendly	Installation cost
Reduces water usage	Storage facilities
Saving on water bills	

We have considered a number of environmentally friendly, sustainable systems and all are designed to reduce our impact upon the environment and reduce running costs. Some of these systems, whilst appearing ideal, are expensive to install and as a result the payback period is often quite long. This tends to make these options appear less attractive, but incentives do exist to encourage their use.

Another consideration is the suitability of the systems for installation in existing properties (retro fitting). The disruption and alterations which may be required can make these impractical as well as extremely expensive.

On new builds and in the case of major refurbishments then serious consideration should be given to the systems we have identified here, finding the most appropriate for the customer and the environment.

One further consideration is the requirement of the local authority with respect to environmental technology schemes such as discussed above. There will be some considerations where, for example, the building is listed or in a conservation area, and different local authorities will have different requirements.

Before planning or carrying out any work it is important to find out what the local authority building control requirements are.

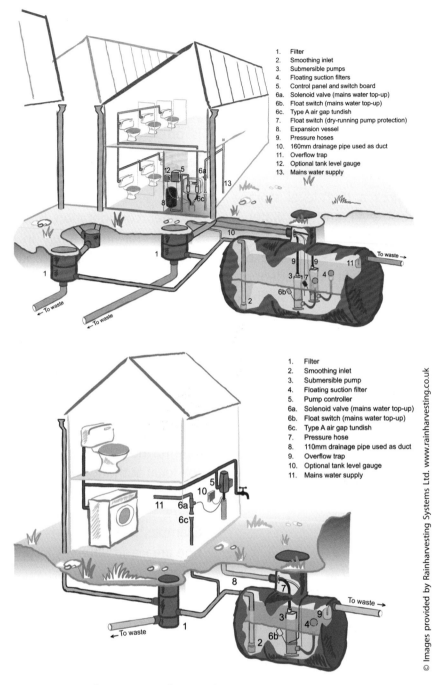

1. Filter
2. Smoothing inlet
3. Submersible pumps
4. Floating suction filters
5. Control panel and switch board
6a. Solenoid valve (mains water top-up)
6b. Float switch (mains water top-up)
6c. Type A air gap tundish
7. Float switch (dry-running pump protection)
8. Expansion vessel
9. Pressure hoses
10. 160mm drainage pipe used as duct
11. Overflow trap
12. Optional tank level gauge
13. Mains water supply

1. Filter
2. Smoothing inlet
3. Submersible pump
4. Floating suction filter
5. Pump controller
6a. Solenoid valve (mains water top-up)
6b. Float switch (mains water top-up)
6c. Type A air gap tundish
7. Pressure hose
8. 110mm drainage pipe used as duct
9. Overflow trap
10. Optional tank level gauge
11. Mains water supply

Figure 2.45 *Typical commercial rainwater harvesting system*

Task

Establish your local authority building control's requirements in regards to:

1   PV systems

2   Wind generators

3   Solar thermal hot water systems.

Try this

1   Solar thermal heating is used to provide:

   a.   electricity
   b.   hot water
   c.   electricity and hot water
   d.   biomass fuel

2   The types of solar thermal collectors are _____ and _____ profile and of these two the _____ profile is more easily incorporated into the roof.

3   Domestic grey water recycling may be used to provide _____ for gardens and water for _____.

4   There are two main types of rainwater harvesting systems which are _____ which requires a high level storage and _____ _____ which may be stored underground.

5   Two of the limitations for rainwater harvesting are the initial installation _____ and the need for a _____ facility.

Congratulations you have now finished this unit and you should take the following self-assessment test which covers all the areas considered in the unit before undertaking the scheme assessment for your particular qualification. Good luck.

SELF ASSESSMENT

1   The output from a PV system may be connected to the public electrical supply system by the use of a:

   a.   rotary converter
   b.   static inverter
   c.   rotary inverter
   d.   static converter

2   The most common type of wind turbine in use today is the:

   a.   vertical shaft
   b.   split shaft
   c.   horizontal shaft
   d.   dynamic shaft

3   In a hydro generator system the water pressure is used to drive a:

   a.   generator
   b.   motor
   c.   dynamo
   d.   turbine

4   Due to the output of a GSHP it is ideal for supplying:

   a.   electricity
   b.   underfloor heating
   c.   all the users' hot water
   d.   ordinary wet central heating

5   One disadvantage of a solar heating system is the:

   a.   low running cost
   b.   high emission
   c.   renewable energy
   d.   long payback time

End test

1. **Part 1 of the Environmental Protection Act sets limits on:**

   ☐ a. water quality

   ☐ b. sewage disposal

   ☐ c. smoke emission

   ☐ d. thermal insulation.

2. **The Pollution Prevention and Control Act regulates pollution from:**

   ☐ a. household waste

   ☐ b. commercial buildings

   ☐ c. site clearance

   ☐ d. industrial processes

3. **One of the factors which determines whether a substance is hazardous waste is if it is:**

   ☐ a. flammable

   ☐ b. chemical

   ☐ c. organic

   ☐ d. metallic

4. **Small household electrical appliances are to be disposed of:**

   ☐ a. in the general landfill waste

   ☐ b. in accordance with the WEEE Regulations

   ☐ c. once they have been dismantled by the user

   ☐ d. without recycling

5. **The disposal and processing of PVC results in the:**

   ☐ a. release of toxins

   ☐ b. production of heat

   ☐ c. release of free radicals

   ☐ d. production of harmless gases.

6. **Hazardous materials must be disposed of by:**

   ☐ a. a general contractor

   ☐ b. an authorized contractor

   ☐ c. an unauthorized contractor

   ☐ d. a jobbing contractor

7. **The Building Regulations provide guidance on the requirements for the:**

   ☐ a. disposal of hazardous waste

   ☐ b. disposal of non-hazardous waste

   ☐ c. further conservation of fuel

   ☐ d. limitation on emissions

8. **Approved Document F in the building regulations relates to:**

   ☐ a. the access and use of buildings

   ☐ b. the passage of sound

   ☐ c. ventilation

   ☐ d. fire

9. The minimum distance from a supporting wall to a hole through a joist is:

☐ a. 0.15 of the joist span

☐ b. 0.20 of the joist span

☐ c. 0.25 of the joist span

☐ d. 0.30 of the joist span

10. The maximum depth of a vertical chase in the 90mm inner leaf of a cavity wall is:

☐ a. 15 mm

☐ b. 20 mm

☐ c. 25 mm

☐ d. 30 mm

11. A downlight installed in the ground floor kitchen of a dwelling must be provided with:

☐ a. a discharge lamp

☐ b. an intumescent hood

☐ c. exposed terminals

☐ d. a plastic bezel

12. A fan installed to provide ventilation for an unvented bathroom may need to have:

☐ a. a run-on facility

☐ b. two points of control

☐ c. a horizontal shaft

☐ d. a separate circuit

13. One method of reducing power consumption for lighting circuits is by fitting lamps which have a luminous efficacy greater than:

☐ a. 40 lumens per watt

☐ b. 30 lumens per watt

☐ c. 20 lumens per watt

☐ d. 10 lumens per watt

14. To meet the requirements of the building regulations the local authority building control would need to be notified if electrical work carried out in a dwelling includes:

☐ a. the replacement of a damaged socket outlet

☐ b. an additional light added in the lounge

☐ c. a new electric shower circuit

☐ d. two additional sockets to a bedroom ring circuit

15. The criteria by which the environmental suitability of new homes are measured are contained in the:

☐ a. Code for Sustainable Homes

☐ b. Building Regulations

☐ c. Environmental Protection Act

☐ d. Environment Act

16. Potable water is also referred to in Approved Document G as:

☐ a. clean water

☐ b. pure water

☐ c. wholesome water

☐ d. safe water

17. Hazardous materials are those which contain hazardous substances in quantities which are liable to cause:

☐ a. unpleasant smells

☐ b. photosynthesis

☐ c. loud noises

☐ d. injury to people

18. Biodegradable materials are:

☐ a. not suitable for recycling but break down harmlessly over time

☐ b. suitable for recycling and do not break down over time

☐ c. not suitable for recycling and do not break down over time

☐ d. suitable for recycling and break down harmlessly over time

19. Adhesive used with plastic conduit is generally classed as a substance which is:

☐ a. non-hazardous

☐ b. hazardous

☐ c. biodegradable

☐ d. inert

20. Waste and damage to materials can be kept to a minimum by installing:

☐ a. all the materials at the beginning of the job

☐ b. materials after coordination with other trades activities

☐ c. electrical material before other trades

☐ d. no materials until other trades are complete

21. Low smoke and fume cables may be installed because they:

☐ a. reduce cost

☐ b. look neater

☐ c. speed up installation

☐ d. reduce environmental impact

22. On discovering an environmental hazard on site it should be immediately reported to:

☐ a. your workmates

☐ b. building control

☐ c. your supervisor

☐ d. a company director

23. Because the output of PV cells is very small to provide a useful voltage they are connected in:

☐ a. parallel

☐ b. delta

☐ c. star

☐ d. series

24. Small PV systems may be used to supply a:

☐ a. dwelling

☐ b. farm

☐ c. road sign

☐ d. shop

25. In order to drive the turbine a wind generator converts wind energy into:

☐ a. electrical energy

☐ b. rotating mechanical energy

☐ c. linear mechanical energy

☐ d. kinetic energy

26. A 'yaw' motor is used in a wind turbine to:

☐ a. turn the blades into the wind

☐ b. turn the turbine

☐ c. generate electricity

☐ d. raise and lower the blades

27. **The ratio between wind speed and energy produced by a wind generator is found from: energy equals:**

☐ a. wind speed

☐ b. wind speed^2

☐ c. wind speed^3

☐ d. wind speed^4

28. **A hydro generator producing an output of 80kW is classed as a:**

☐ a. mega-hydro system

☐ b. micro-hydro system

☐ c. hydro system

☐ d. pico-hydro system

29. **Where a simple waterwheel is used to generate electricity a gearbox would be required because of the:**

☐ a. slow speed of the wheel

☐ b. high speed of the wheel

☐ c. variation in water speed

☐ d. changes in water pressure

30. **A GSHP derives low grade heat from the ground and transfers this to the installation by way of:**

☐ a. a converter and generator

☐ b. an evaporator and compressor

☐ c. a compressor and converter

☐ d. an evaporator and converter

31. **A heat pump produces:**

☐ a. electricity

☐ b. warm air

☐ c. gas

☐ d. hot water

32. **The figure shown above is for a typical:**

☐ a. Ground source heat pump system

☐ b. Air source heat pump system

☐ c. Combined heating and power system

☐ d. Hydro generation system

33. **MicroCHP systems generally have an output in the region of:**

☐ a. 1kW

☐ b. 2kW

☐ c. 5kW

☐ d. 10kW

34. **A heating system which derives its source fuel from landfill sites is known as:**

☐ a. biomass

☐ b. bioliquid

☐ c. biofuel

☐ d. biogas

35. **Common UK biomass fuels are coppiced willow and:**

☐ a. oak

☐ b. ash

☐ c. poplar

☐ d. chestnut

36. **Solar heating systems are used to supply hot water produced using heat from the sun onto collectors which are either flat profile or:**

☐ a. square profile

☐ b. evacuated tube

☐ c. triangular profile

☐ d. pressurized tube

37. **Grey water harvesting uses water from domestic use excluding:**

☐ a. showers

☐ b. toilets

☐ c. laundry

☐ d. bathing

38. **The water collected from rainwater harvesting is most commonly used for:**

☐ a. irrigation

☐ b. drinking

☐ c. cooking

☐ d. bathing

39. **Installing environmental technology systems in existing premises requires consultation, before installation, with the:**

☐ a. health and safety executive

☐ b. local authority

☐ c. water utility company

☐ d. local residents

40. **The grey water waste produced in residences represents what percentage of the total water consumption?**

☐ a. between 20% and 40%

☐ b. between 30% and 50%

☐ c. between 40% and 70%

☐ d. between 50% and 80%

Answer section

Chapter 1.1

Try this Page 6

✓

✓

☒

☒

✓

☒

Try this Page 9

Everyone who is at work.

To observe safe working practices.

The standards of Health, Safety and Welfare of all those at work.

adequate ventilation
fume and dust control
adequate lighting
clean and tidy workplace.

washing facilities
sanitation
first aid facilities
safe method of work
safe storage, handling and transporting of goods
training.

hand protection
foot protection
a safety harness or belt.

The Health and Safety Executive (HSE) or the Local Authority Environmental Health Department

Accident book.
Improvement Notice
Prohibition Notice.

Try this: Wordsearch Page 13

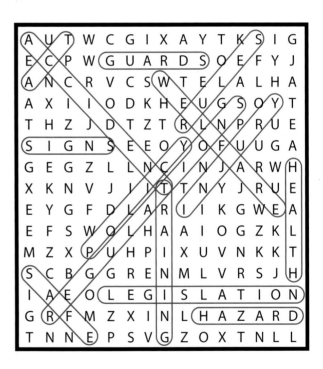

SELF ASSESSMENT Page 16

1. c) both employers and employees
2. c) safety training
3. a) white cross on a green background
4. d) hazards present in the workplace
5. d) hand tools for installation work

Chapter 1.2

Recap Page 17

Everyone at work

1) Reasonable working temperature
2) Reasonable working humidity
3) Adequate ventilation
4) Fume and dust control
5) Suitable and adequate lighting
6) Clean and tidy workplace

The accident book and to the Health and Safety Executive or Local Authority Environmental Health Dept.

Try this Page 21

circuit protective conductor

reduced low   safety

Residual current device

BSEN 60309-2:1992   interlocking   different positions

Try this: Crossword Page 26

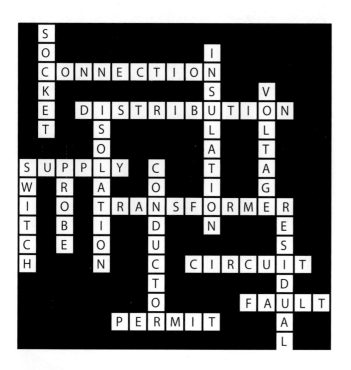

SELF ASSESSMENT Page 27

1. c) 30mA
2. a) blue
3. c) 55V
4. d) Class II equipment
5. a) isolated and locked off

Chapter 1.3

Recap Page 28

visually

circuit protective conductor

double insulated

isolator

GS 38

permit   work

Try this Page 35

Take care not to become a casualty yourself.

Fuel, air and a source of ignition.

People smoking carelessly, friction heat, sparks, naked flames, fuel leaks, faults or failures in equipment or any other suitable answer.

Water or foam based extinguisher.

Water extinguisher with red label.

SELF ASSESSMENT Page 36

1. b) disconnect the supply
2. a) fuel, air and a source of ignition
3. d) dry powder   LPG
4. b) receive a shock
5. b) flush the burn with cold water

Chapter 1.4

Recap Page 37

connection

dry

fuel, air and heat

water   foam based

Fire blanket

Try this Page 44

barriers and warning notices

gas, electricity   compressed air   bonded

Try this: Crossword Page 53

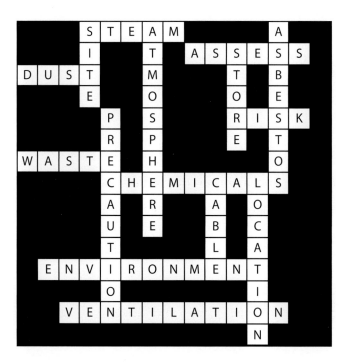

SELF ASSESSMENT Page 54

1. c) 18 years
2. c) raised off the floor

3. a) a delivery note
4. d) passes through a structural brick wall
5. b) a lamp crusher

Chapter 1.5

Recap Page 55

hazard

Insecure structures, inadequate lighting, risk of falling, risk of being hit by falling objects, risk of drowning, dangerous or unhealthy atmospheres, steam, smoke or vapours…

A locked hut or room, not damp, materials and equipment laid out so no damage can be caused to it.

Control of Substances Hazardous to Health

cancer   lung

Try this: Crossword Page 60

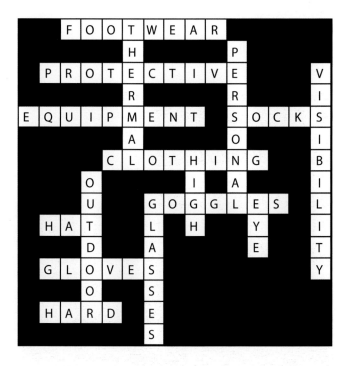

SELF ASSESSMENT Page 65

1. d) white gloves on a blue background
2. b) triangular with a black border
3. a) COSHH
4. d) shows what must not be done
5. d) a white telephone on a green background

Chapter 1.6

Recap Page 66

Eye protection in the form of goggles or faceshield, body protection and hand protection.

falling   flying

Mandatory   circular   white   blue

red

Task Page 68

Task, Individual, Load and Environment

Steep slopes and rough surfaces

Lifting Operations and Lifting Equipment Regulations 1998

thoroughly examined by a competent person

Try this Page 70

7kg

10kg

equipment   trained

Try this Page 73

necessary

Weight, bulk, difficult shape, greasy, sharp corners, hot

Obstacles on the way to or at the destination, headroom, fixtures in the way, uneven or slippery floor, bulky light load that could be caught by the wind, level of illumination

capable

load   twisting

reduce   injury   stage

back kept straight

let legs do pushing

Try this Page 76

$$\frac{12 \times 0.6}{0.3} = 24\text{kg}$$

Try this Page 79

$15 \times 2 = 30\text{kg}$

$$\frac{36}{4} = 9\text{kg}$$

$$\frac{72}{18} = 4$$

= 4 pulley system

Try this Page 80

20kg   mechanical

Wheelbarrow, sack barrow, flat trolley, table trolley

slings   pulleys

moved

SELF ASSESSMENT Page 80

1. d) bent legs and upright back
2. d) 8kg
3. a) 10kg
4. b) shift a load to another location at a similar height
5. c) 20kg

Chapter 1.7

Recap Page 81

Block and tackle, winch

Electric motor

Hydraulic lift

Petrol engine, diesel engine

Try this Page 87

Answer relates to individual

Ladder

Step up

A pair of steps

5 rungs or 1.05m

Try this Page 90

$$1.4m \left(\frac{5.6}{4} = 1.4m \right)$$

Try this Page 93

steps   ladder

trestles   scaffold

assembly   alteration   competent person

4:1   75°   5

firm   good

not   electrical equipment   short   electrical shock

SELF ASSESSMENT Page 94

1. c) 5
2. b) 30 minutes
3. c) 4:1
4. d) a competent person
5. b) 2

End test

1. a. Statutory document
2. d. Provides guidance
3. c. Unable to work for more than three days due to the accident
4. b. An HSE inspector
5. c. Setting a given time to achieve the improvement

6. a. Control of substances hazardous to health
7. d. Five employees
8. c. Provision and Use of Work Equipment Regulations.
9. b. No mobile phones
10. c. First aid box
11. d. White gloves on a blue background
12. b. Triangular with a black border
13. c. CO_2
14. c. Enable steps to be taken to prevent recurrence
15. c. Electricity at Work Regulations
16. d. Report it to a supervisor immediately
17. a. A risk exists which cannot be removed
18. b. Eye protection
19. c. Any danger warnings
20. b. Workplace (Health and Safety and Welfare) Regulations
21. d. Stop work immediately and report it
22. c. Dermatitis
23. a. Suitable procedures are followed
24. c. Carbon dioxide
25. c. Perform mouth to mouth and call for help
26. c. Reduced low voltage
27. c. Details of the work to be carried out
28. b. No work other than that identified in the permit
29. d. To provide employee's hand tools
30. c. Lifting gear
31. b. 12kg $\frac{24 \times 0.3}{0.6} = 12kg$
32. a. Safe working load
33. d. 2kg
34. c. 30 minutes
35. c. 75°
36. a. Mobile access towers
37. d. 470mm
38. b. 18
39. c. 7 days
40. b. Electric shock

Chapter 2.1

Try this: Wordsearch Page 108

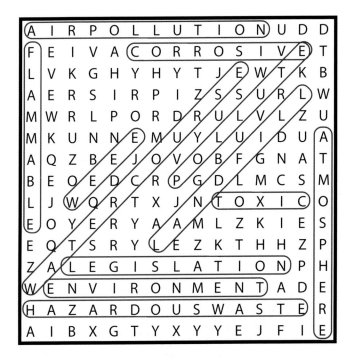

SELF ASSESSMENT Page 109

1. b) Environmental Protection Act
2. c) local authority
3. c) corrosive
4. d) 1000V ac
5. c) land and water courses

Chapter 2.2

Recap Page 110

Pollution Prevention and Control Act (PPCA)

The Waste Electrical and Electronic Directive (WEEE)

Air Pollution, Land Pollution and Water Pollution

Try this Page 113

1. B
   F
   P
2. a) Internal fire/intumescent pad
   b) External fire/intumescent pad
3. b) 15mm
4. c) 1.50m
5. b) 2

Try this: Crossword Page 118

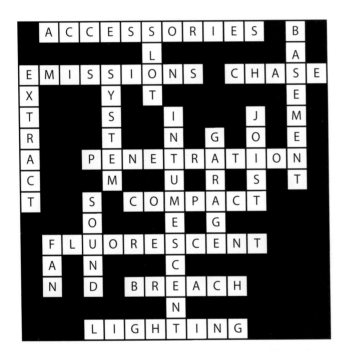

SELF ASSESSMENT Page 124

1. c) All new dwellings
2. d) 450mm
3. b) Dwelling
4. d) The installation of a new circuit supplying a cooker
5. a) drinking water

Chapter 2.3

Recap Page 125

Approved Document B, Volume 2

Storage area

Automatically switched off when there is sufficient daylight and when it is not required at night

450mm

No it is not notifiable

SELF ASSESSMENT Page 131

1. b) asbestos
2. a) paper
3. c) well ventilated
4. d) mercury
5. b) up to the point of disposal

Chapter 2.4

Recap Page 132

i)   Death or injury to people and wildlife
ii) Pollution of waters
iii) Pollution of air
iv) Unacceptable impact on the environment if not properly dealt with.

Asbestos

hazard   recycled

hazardous   COSHH

non-hazardous

types   registered

SELF ASSESSMENT Page 137

1. b) careful planning
2. c) delivered as they are required
3. a) the appropriate ambient temperature
4. d) immediately
5. b) low smoke and fume

Chapter 2.5

Recap Page 138

reduce   installation

delivered   minimizes

reported promptly

improved   environmentally   products

impact   reduction   chemicals

Try this Page 142

series   small   dc   static inverter

blades   vane   yaw motor

horizontal shaft   vertical shaft

Try this Page 145

turbine   reservoir   flowing

Forebay
Penstock
Turbine
Generator

ground   air   source   building   refrigerator

d) 300%

Try this Page 148

c) combined heat and power

heat   electrical   heating   water

b) 1.0kW

Try this Page 153

b) hot water

flat profile   round   flat

irrigation   flushing toilets

header tank   direct pump

cost   storage

SELF ASSESSMENT Page 153

1. b) static inverter
2. c) horizontal shaft
3. d) turbine
4. b) underfloor heating
5. d) long payback time

End test

1. c. smoke emission
2. d. industrial processes
3. a. flammable
4. b. in accordance with the WEEE Regulations
5. a. release of toxins
6. b. an authorized contractor
7. c. further conservation of fuel
8. c. ventilation

9. c. 0.25 of the joist span
10. d. 30mm
11. b. an intumescent hood
12. a. a run-on facility
13. a. 40 lumens per watt
14. c. a new electric shower circuit
15. a. Code for Sustainable Homes
16. c. wholesome water
17. d. injury to people
18. a. not suitable for recycling but break down harmlessly over time
19. b. hazardous
20. b. materials after coordination with other trades activities
21. d. reduce environmental impact
22. c. your supervisor
23. d. series

24. c. road sign
25. b. rotating mechanical energy
26. a. turn the blades into the wind
27. c. wind speed^3
28. b. micro-hydro system
29. a. slow speed of the wheel
30. b. an evaporator and compressor
31. d. hot water
32. c. Combined heating and power system
33. a. 1kW
34. d. biogas
35. c. poplar
36. b. evacuated tube
37. b. toilets
38. a. irrigation
39. b. local authority
40. d. between 50% and 80%

Glossary

ASHP Air source heat pump

BER Building Emission Rate (carbon dioxide)
BS7671 Requirements for Electrical Installations
BSI The British Standards Institution

CFC chlorofluorocarbon
CHIP Chemicals (hazard information and packaging for supply) 2009 regulations
CHP combined heat and power
Class II equipment often referred to as double insulated equipment
Contaminant Something that pollutes the environment
COSSH Control of Substances Hazardous to Health
CPC Circuit protective conductor

Danger exposure to harm
DER Dwelling Emission Rate (carbon dioxide)

EAWR Electricity at Work Regulations
Emergency a sudden crisis that must be dealt with quickly
Environment surrounding influences

Fulcrum the point or support about which a lever turns

GSHP Ground source heat pump

HASAW The Health & Safety at Work Regulations
Hazard Something that could potentially be harmful to a person's life, health, property or the environment.
HCFC Hydrochlorofluorocarbons
HSE The Health and Safety Executive
Hydraulic a device driven byliquids

INDG Industry guidance
Irritant causing irritation

LED Light emitting diode
LOLER Lifting Operations and Lifting Equipment Regulations 1998
LSF Low smoke and fume

Mandatory means YOU MUST DO.
Method statement A method statement details precisely what is to be done, how it is to be done, any special requirements or actions and the time anticipated for the work to be carried out.
MEWP Mobile elevating work platforms
microCHP micro combined heat and power

noxious harmful or injurious to person's health or well-being

PASMA Prefabricated Access Suppliers' and Manufacturers' Association
Permit to work a permit to work scheme ensures that activities in areas of risk are controlled and monitored to minimise danger
Pneumatic a device driven bycompressed air
Pollution where contaminants have been introduced into a natural environment
PPCA Pollution Prevention and Control Act
PPE Personal protective equipment
PUWER Provision and Use of Work Equipment Regulations 1988 (PUWER)
PV photovoltaic

RCD residual current device
Risk the chance that injury or damage will occur
RoHS Restriction of Hazardous Substances

SWL Safe Working Load

TER Target carbon dioxide Emission Rate
TILE Task, Individual, Load and Environment

WEEE The Waste Electrical and Electronic Equipment Directive

Index: Electrical Installation